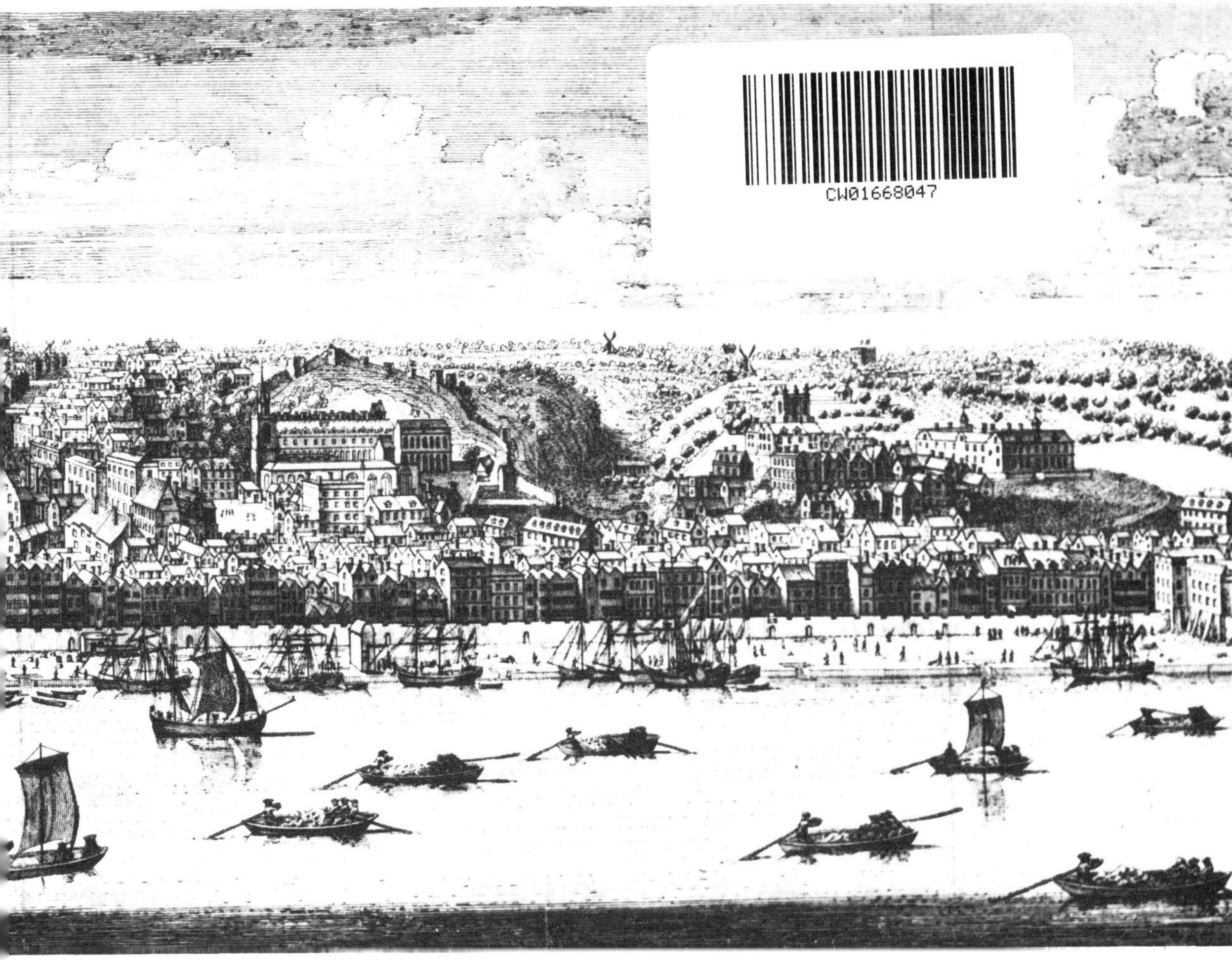

veen 1727 and 1753. It shows the keels, sail and rowing, with coasting hoys taking the tide down to the sea.

1850

COALS FROM NEWCASTLE

COALS FROM NEWCASTLE

The Story of the North East
Coal Trade in The Days of Sail

by

Roger Finch

TERENCE DALTON LIMITED
LAVENHAM . SUFFOLK
1973

Published by
TERENCE DALTON LIMITED
SBN 900963 39 5

Chapter headings are from wood engravings by Thomas and John Bewick both working in Newcastle c 1780–1828. Chapter 11 heading is by Roger Finch.

Printed in Great Britain at
THE LAVENHAM PRESS LIMITED
LAVENHAM SUFFOLK

Contents

Index of Illustrations

Introduction and Acknowledgements

THE mining of the Great Northern coalfield helped to initiate our country's industrial revolution and profoundly shaped the face and social structure of a whole region. Its exploitation upon an unprecedented scale in the seventeenth and eighteenth century was almost entirely dependent upon an ever expanding market in London and the South East, made accessible by sea and river, a watery highroad which stretched from the pit-head to the consumer. From medieval days 'sea-cole' was brought south by ever-growing fleets of sailing-colliers; fleets which were the wonder and astonishment of contemporary diarists. As the trade expanded so did the wealth of Tyneside, Wear and later Tees. The confidence engendered by successful maritime-based enterprise and the growth of trade primed the pump for later, more complex industrial development of which ship-building continued as a primary activity. The wooden sailing-ships in the coal-trade voyaged to the Baltic and the Americas, across the North Sea, to Spain and France; but it was the coastwise trade which brought the fleets into being and it was coal which provided cargoes whose total tonnage exceeded all others by many thousands of tons.

The carriers of 'sea-cole' emerged as Britain's first group of specialised sailing vessels. This emergence probably took place in the early seventeenth century and they remained distinctive in build and rig until the end. The fleet survived into the twentieth century, but less gloriously, as a group of sailing vessels built originally for ocean trading, surviving precariously upon the freights despised by the steamers, trudging the length of the North Sea. With worn canvas, minimal crews and their pumps in constant use, the last of the sailing colliers could be seen until the First World War. Even as late as the nineteen fifties it was still possible to find a coasting spritsail barge discharging coal at an East Anglian quayside, lying in a mud berth which had cradled the hulls of sailing-colliers ever since the first hoy had arrived, five weeks out from Tynemouth, in Tudor times.

We shall be primarily concerned here with the sailing fleets of pre-steamer days, their owners and builders; how they loaded, navigated the treacherous sea-way to London, how they discharged their cargoes and the organisation of a trade which was beset with so many archaic restrictions. The later history, the history of the powered collier must be left to some-one more devoted to steam, steel and the sophisticated economics which supported them than myself. This is a story of a time when no creek was complete or harbour properly furnished without the heavy topmast and blackened rigging of a Geordie brig rising above the wooden warehouses and when the lower Thames was clogged with fleets, numbering scores, drawn from the north eastern coal ports. The men, too, who commanded and manned them deserve full recognition. When one learns of the conditions under which they laboured and the dangers they faced one can only echo the sentiments of a Parliamentary report of 1800. . .

> "The difficulties of the navigation of the Coal Trade give the seamen derived from it, in point of skill, expertness, patience of fatigue, and hardship an incontestable superiority over those drawn from other maritime trades".

The existence of a great sailing collier fleet had important effects upon maritime developments in many spheres. Change only came slowly, but the initiation and growth of the life-boat service was directly related to attempts to alleviate the desperate plight of sailing colliers in distress. The early provision of navigational aids, buoys, lighthouses and lightships along the East coast of Britain similarly led the world. Without the incentive provided by the coal-trade fleet these services would never have become widely based so early or so well maintained. The first effective application of steam towage ever devised was to handle the wind-bound coal carriers.

Even in decay the sailing fleet had its effect. The abuses that the management of the worn sailing colliers in some cases permitted, finally brought about the framework of legislation which regulates shipping to-day and was largely initiated from Tyneside rather than Whitehall.

We are now witnessing a sharp contraction of this seven centuries-old seaborne trade in coal. The headlong expansion of the Northern coal fields, which marked the middle decades of the nineteenth century, has been replaced in our day by a more orderly and infinitely more humane contraction. The export trade has gone completely. No longer can one stand on any North Sea headland or South Coast cliff-top and see, as for so long past one could have seen, a loaded collier; a collier dipping her fo'c'sle under the swell, water running from her scuppers, outward bound, or yawing and pounding high in the water, light loaded for Blyth. More easily won fuels are replacing coal. It is perhaps a fitting moment to record the sailing collier which pioneered such an important trade with such far-reaching consequences.

The most pleasant part of writing this book has undoubtedly been the opportunities it has afforded me to revive old friendships, enjoy new ones and unashamedly take advantage of the erudition of a host of acquaintances who share an enthusiasm for ships and the sea. Without the initial encouragement of John Lewis and the expert guidance of Hervey Benham the completion of the task would have been much less enjoyable. Twenty five years of talking history and boats with Hugh Moffat have not exhausted his deep understanding of the subjects, while Peter Barton brought an invaluable north-countryman's critical ability and an immense and unstinting practical assistance. I should also like to acknowledge the help of Basil Greenhill, Director of the National Maritime

Museum, Colin Sorensen of the London Museum, and Gilbert Pleuger. Without the aid willingly provided by the R.N.L.I., the National Coal Board, Frank Atkinson of the North Regional Open Air Museum, Beamish Park and Robert Malster the illustrations to the book would be far less comprehensive.

Finally I am particularly indebted to Marjorie Smith for her inexhaustible care in the preparation and checking of the manuscript and ensuring its final appearance in a comprehensible form.

Roger Finch,
Ipswich, Suffolk.

September, 1973.

Sailing Collier 1650. *Roger Finch*

I

The Earliest Days

ON A grey April morning in 1904, in thick weather the brig *Remembrance* of Whitby, deep loaded with sea-coal and five days out from Hartlepool, was beating against a S.W. gale up the inshore channel off the Suffolk coast between the Whiting and Bawdsey banks. But between the sands lie the rocks of the Kettle Bottom, unmarked and here the little wooden vessel struck, bumping several times, rising and falling to the wind-over-tide swell and then driving on into deeper water. The ageing hull rapidly began to take water far faster than the crew could work the primitive iron pumps. Aloft the grey canvas was given a rough stow, but there was little else that could be done. It was an experience that was considered almost part of the daily round of a coasting seaman under sail. The constant danger of shifting sands and unheralded changes of weather, of steamers pounding on their way with slight regard for sail and failures of spars aloft and old wooden hulls below, gave these seamen in sail an almost fatalistic attitude to such disaster.

With the sullen rise and fall of the brig's hull to the waves becoming ever more leaden it became clear that the vessel was beyond assistance. There was a hasty conference between the master and mate and the decision to abandon ship was reluctantly made. No doubt, they agreed, one of the watchers at the coast-guard station at Orford Haven had seen the stricken vessel and a message to alert the life-boat had been despatched. But the shore was three miles distant across grey, tumbling waves and a cruel line of broken water separated the crew of the collier from safety. Clothing, belongings and documents were hastily gathered together, a water breaker and a few meagre stores stowed in the two ship's boats which were then hoisted out to leeward where the floundering hull of the brig offered some protection. The crew, captain, mate and six hands scrambled over the rail, the skipper in charge of one boat, the mate commanding the other. No especially equipped lifeboats these, but heavily built of oak and elm to withstand punishing work in crowded docks and in an emergency such as this, man enough for the job – just. The crew prevented the surge of the waves from dashing them along scarred wooden sides of the *Remembrance* with a skill learnt of long experience and then pulled clear of the wreck. Fully concerned with the task of reaching safety they gave the brig only the briefest of backward glances as with sails flagging in the wind and yards awry she had soon only her topgallant masts above the waves.

But their predicament had not gone un-noticed by the Coastguards. The Aldeburgh lifeboat *City of Winchester* had been launched, as it had been so often in the past to render assistance to sailing colliers in distress and was beating towards the wreck under its powerful brown lugsails. A hail in the waste of tumbling water to a passing smack and the coxswain knew their task was over. The Aldeburgh cod-smack *Gipsy,* fishing off the Suffolk shore had been a witness to the tragedy and had forestalled their rescue attempt. Being up wind of the wreck the fishermen had rapidly reached the survivors. The boats were made fast to the taffrail of the smack and with the exchange of a few laconic remarks the castaways had scrambled aboard the heaving deck. With the sails sheeted hard in and the rum bottle circulating round the cabin stove the *Gipsy* turned for Harwich. By a strange quirk of fate the last of the cod-smacks had rescued the crew of the last collier brig. Ten days later the wreck of the *Remembrance,* a danger to navigation, lying as she did in the channel between the Whiting and Shipwash, was blown-up by charges placed by the Trinity House tender and so, as her cargo split across the sands from her broken hull, the last of the Geordie collier-brigs, which had numbered hundreds in their day, met her end.

The *Remembrance,* at the time of her end an anachronism amongst the still numerous fleet of sailing coasters, was a worthy survivor from a vanished fleet. In her varied career of nearly half a century she had been associated with many of the coal-ports, both old and new, whose story concerns these pages. The form of her hull and rig, incongruous even at the time of her loss, preserved much of the appearance of the colliers two hundred years before, when they were the pride of the coasting fleet. On what was to be her final voyage destined for Greenwich, she was loaded with coals from Hartlepool, a coal-port on the northern extremity of the Tees peninsular on whose southern shore she had been built. Although at the time the *Remembrance* was launched from Craggs' yard at Middlesbrough in 1865, the iron, steam-powered collier was an established fact, she was built to join a fleet of similar sailing vessels still hundreds strong. Indeed the last traditional collier-brig, the *Ocean Belle* was not launched until 1874, at Blyth. Like many of her contemporaries whose tarred hulls and long bowsprits were as well known in the Northern European ports as on the London river, the *Remembrance's* first voyage was with Sunderland coals to the Baltic. Originally registered at Whitby and owned there, the brig later passed into the numerous fleet assembled by James Knott of Quayside, Newcastle, in the eighteen eighties, one of a mixed collection of steamers and ageing sailing colliers shrewdly accumulated during the transition from canvas to steam and then immediately disposed of without sentiment when their maintenance became uneconomic. Sold from the Tyne the *Remembrance,* still staunch enough to be given a Lloyds' rating, found an owner at West Hartlepool, then a few years later, another at Sunderland and finally returned, appropriately to Whitby, there to be owned by G. Moorsom. At Whitby there still lingered around its harbour an atmosphere of wooden ships,

The *Remembrance*, last of the North East collier brigs at Gravesend. Launched at Middlesbrough in 1865 and lost in 1904. *National Maritime Museum, London*

the craftsmanship which served them, and in the 'Haven under the Hill' there gathered, among the smacks and cobles, the last remnants of the sailing 'sea-coal' fleet redolent of the time before iron, steam and grime had transformed the North East ports and their rivers.

'Sea-coal' was a name commonly used by producers, shippers and consumers until as late as the twentieth century but it had been used since early medieval times. However, we must not state too unequivocally that 'sea-coal' derived its name from the route by which it reached London merchants. Perhaps the name was derived from the out-cropping of coal measures on the seashore of Northumberland where it could be easily gathered from the tideline without recourse to even the most primitive mining. This was the fuel the Romans had used for smelting and domestic warmth; they heated the hypocausts beneath their villas with it and indeed bunkers of coal have been found in forts on or near Hadrian's Wall. The Latin word *carbo* signifies both of coal and charcoal, so that in the earliest documents it is not easy to determine which material is referred to.

Perhaps 'sea-coal' was the distinctive name used to distinguish the mineral from charcoal, for coal denoted what we now call charcoal in the Northern Counties in Saxon times. But for the merchants and factors who handled the cargoes delivered by the *Remembrance* and her countless ancestors 'sea-coal' implied fuel mined from Tyneside and the Wear and shipped to the London market.

Much of our information on the early centuries of the North East coal trade inevitably derives from documents relating to legislation and restrictions or ordinances limiting its use and of orders taxing it when it was used. Very early coal was recognised as a precursor of a radically different way of life with the result that from the thirteenth century onwards it was hedged with laws and regulations. In a world which was largely without drains and usually unwashed, the medieval lords and ladies were, if one may judge from the frequent complaints they voiced, unduly susceptible to coal smoke. Perhaps as chimneys were by no means universal, they are mentioned as a new phenomenon in Norman times, their objections are understandable.

The early use of coal as an industrial fuel left its mark on the topography of the City of London where a Sea-Cole Lane, also known as Lime-Burner's Lane, existed in the thirteenth century. This was conveniently placed for the landing of cargoes from the nearby River Fleet. Before it was ignominiously banished underground to serve as a sewer this was a navigable creek of the Thames and survived as such until the time of Pepys. However, as early as 1285 complaints were lodged that the smoke from the kilns of the Fleet-side lime-burners was causing a nuisance, perhaps the earliest known objection to industrial pollution and as ineffective as many since, for the objection was raised again three years later. In 1305 the use of coal was completely banned, although completely unsuccessfully. The earliest evidence we have is of coal being used for industrial purposes rather than domestic heating. The building and reconstruction of London throughout the centuries created an insatiable demand for fuel brought by sea required for lime manufacture, while the forges of the smiths and armourers called for a temperature and a flexibility of heat which locally produced charcoal could not provide. The smiths, too, were reprimanded for causing "undue blackening of the air" when they employed sea-coal and in 1298 the master-smiths promised to refrain from working by night. No doubt the near-neighbours of these craftsmen must have been as thankful as the more influential, though more distant, petitioners against smoke. Noise as well as soot were no doubt capable of polluting the environment in the Middle Ages as they are in the twentieth century.

The quickening of commerce and technology which has recently been recognised as marking the twelfth and thirteenth centuries extended to the exploitation of minerals for the use of emergent industries. By the end of the thirteenth century archaeological and documentary evidence points to the mining of coal, or perhaps "the extraction of coal" defines the process better, for the skills of the industry were as yet on a primitive level in all the coal-fields of Britain. The exploitation of the great majority of coal measures was severely limited by the lack of any effective form of transportation; distribution was restricted to an area little more than a dozen miles in radius from the point of extraction. Horse-drawn wagons sank axle deep in the mire of roads scarcely distinguishable from trackways. The primitive drift-mines were worked by peasants in the interval between harvest and winter, by which time haulage became impossible with churned-up roads and beasts weak from lack of fodder. Furthermore there were legal difficulties, for the severity of the medieval Forest Laws prevented more than a surreptitious scratching of the 'waste' even when logistic problems could be overcome. It required the royal authority of Henry III to be given before the merchants of Newcastle were allowed to dig for coal and to exploit their uniquely favourable position on the Tyne.

In three geographical areas the constricting limitations of land transport did not apply. As we have seen, where carboniferous strata are exposed on the seashore or river bank, a flexible means of distribution was at hand. Once collected, even only as a possible buffer against future hardship, coal could await shipment without deterioration. It provided an easily stowed cargo, unaffected by salt water and everywhere in demand; ideal for the early merchant ships whose eventual destination was uncertain, whose time of arrival was problematical, and whose holds were far from watertight.

Conditions of easy extraction, linked with potential water transport, obtained in three clearly defined areas. As well as on the Tyne and Northumberland seaboard, coal measures and navigable estuaries coincided on the shores of the Firth

The wooden ships in the coal trade required constant attention, repair and caulking. Here the *Margaret Nixon* is dry-docked at Whitby; she was one of the last brigs built at Newcastle, launched in 1861, 188 tons reg.

Sutcliffe Gallery

of Forth. As early as 1214, at Carriden, near Blackness of the Firth of Forth, monks of Newbattle Abbey were granted rights to gather coal on the sea shore at "Preston, in the land of Tarent". There were on the west coast, in South Wales, similar juxtapositions of mineral wealth and estuary conditions.

Historical accident and geological factors explain the growth to pre-eminence of the North East coal field amongst these three areas. The more northerly mineral wealth of Scotland was, like the Tyneside region, developed in conjunction with a seagoing fleet of colliers, but during the rapid Tudor expansion which the Tyne enjoyed, Scotland was, as far as the markets of London and the south were concerned, a remote foreign country, with which it was difficult to establish trading links. Anglo-Scottish relations were stormy and, until a much later date, did little to assist a connection between producer and southern consumer. By the time political relations had stabilised the market had been secured by a rival who was three days sail nearer to London. On the other coast, to the west, the primitive seashore mines of South Wales were too distant to form an effective competitor for the lucrative markets of the south eastern seaboard; they lacked at this time an incentive to stimulate production beyond a level that satisfied merely local requirements.

One of the earliest surviving references to coal production in the Northumberland area comes from a document relating to Newminster Abbey, near Morpeth. In 1236 the community there, close by the river Wansbeck, was granted from the King, land on the seashore with right of way "to gather sea-weed for tillage and sea-coal where-ever it might be found"; a nice balance of good stewardship. However, the austere monastic rule stipulated that only in the 'warming house' might the monks light fires for other purposes than cooking. We must hope that even tempted by the bitingly cold northerly gales they did not misuse their bounty. Certainly four years later another Charter specifically conceded that the coal was to be used for the Abbey's forge; monasteries with their pioneering drive and advantage of continuity did much in the furtherance of husbandry and medieval technology. But such an expansionist policy inevitably led to collisions with the lay authorities. At Newcastle in 1268 a case was brought to the justices involving a complaint about the taking away of a ship of the Prior of Tynemouth, lying laden with sea-coal. This is unlikely to imply a case of straightforward municipal piracy, but rather that the burgesses of Newcastle intended to enforce their rights of navigation of the Tyne and levy various dues imposed upon vessels and cargoes belonging to strangers, from which not even the Prior should be exempt. The record of this legal action is one of the earliest pieces of evidence we have linking the city fathers of Newcastle with their long-fought battle to maintain an iron grip over the winning and distribution of Tyneside coal. The villages which were later to become the coal ports of North and South Shields had been founded in the face of opposition from Newcastle by local ecclesiastical houses and this

In 1828 E. W. Cooke drew and etched this collier brig discharging her cargo for his book *Shipping and Craft*. Nearly a century later when the *Remembrance* was lost they had changed but little.

made Newcastle's burgesses even more determined to prevent any further encroachment, however much such development might be justified by divine inspiration.

But we are not primarily concerned here with the indignation of burgesses in the face of pushing Priors. It is this, our first reference to a seagoing collier which is of paramount interest. Specialisation of purpose in shipping was scarcely known until modern times; warships were hastily converted merchant vessels until the introduction of seagoing artillery in the time of Henry VIII's reign. While specialised tackle and equipment were shipped for particular voyages, hulls and rigging showed little differentiation. We must bear in mind, therefore, that to refer to the Prior's vessel as a collier is to state she was a coaster amongst whose varied cargoes coal was one. But as at this date the more lucrative carrying trades were handled by the *cogs* and *nefs* of the Hanseatic traders, the less glamorous 'dirty' cargoes were probably found in such a vessel's hold.

To reconstruct the appearance of this early coal-carrier is difficult. We are reconstructing a coastal trader of Northern waters, launched nine centuries ago, capable of carrying bulky cargoes, of crossing Tyne Bar and lying in the Prior's Haven at Tynemouth Priory. The harsh tides of the North East Coast have left few, if any, wrecks to be discovered by the marine archaeologist; hulks that were abandoned did not, as they might have done further south, find a resting place in soft estuarine mud, to await discovery by chance industrial excavation in our own time. Abandoned upon open beaches at the end of their useful working life, they were pillaged for timber and for their metal fittings. However, on the shores and fiords of the tideless Baltic and the drained area of the Zuiderzee, both areas which shared a tradition of primitive shipbuilding similar to that of the North East coast, numerous ship-finds have been made. These have made possible the tentative reconstruction of the ship development from earliest times in the area. The reconstruction of physical remains has been assisted by other means. Ships motifs were used on the seals of many seaports, from Bergen to La Rochelle, from the second half of the thirteenth century onwards and they appear in medieval church frescoes. These provide a valuable source of evidence upon which exhaustive study has been focused in the past and may now be related to the actual remains of the vessels.

To relate finds from opposite sides of the North Sea in order to reconstruct a Northumbrian vessel might be thought questionable. However, the contacts at every level were commonplace. Newcastle at this time could command more than ordinary resources of skill and material, for one of a series of great galleys ordered by Henry III was built at the port. The extent of the sharing of shipbuilding traditions with Scandinavia may be judged from the very close relationship between the English terminology relating to the galley's construction and Old Norse. Collating these various sources of evidence we can imaginatively reconstruct the earliest recorded coal-carrier as a single masted vessel, sharp-sterned, clinker built, and steered with an oar. The construction of quite large clinker-built vessels (those with each plank of its outer skin overlapping its lower neighbour) was univeral in Northern waters until post-medieval times and survived until the nineteenth century to be used in quite large merchant vessels. It required the minimum of calculation to construct and produced an easily driven, simply repaired hull; however, beyond a length of some 100 feet it was unsuitable for carrying dead-weight cargoes to sea. But this form of construction was abandoned only with reluctance as more sophisticated methods spread from the south. Such an early medieval collier would be scarcely capable of working to windward and groped its leaky way along an uncharted coast with only the most primitive navigational equipment. Reliance was placed upon the three 'L's' traditional to the

East Coast – lead, log-line, and look-out. The compass was not found in general use aboard such workaday coastal craft for another two centuries and was only to be found on large ocean-voyaging vessels in a primitive form; charts were non-existent. Much voyaging consisted of hastily snatched passages, between havens which gave little protection, made perforce between sunrise and sunset and rarely covering more than 70 miles.

Much of a medieval seaman's life must have been spent crouched in the lee of the weather bulwark while his gruel slowly boiled in the vessel's single, communal iron cooking vessel, heated on a stone hearth on deck. No doubt our early collier's crew improved their lot by bartering 'small coals' for fish from the longshore craft, just as their successors were to do until the end of the trade under sail six centuries later. The fact that so small a vessel as these medieval traders was capable of beaching without undue hazard enabled temporary repairs to the simply constructed hull to be effected between ports and an error in navigation might be made without catastrophe being inevitable. But these advantages were minor compensations for the complexities of working along a low-lying coast whose only navigational aids were the cressets burning from the towers of coastal monasteries as at Whitby, and churches such as Blakeney.

Yet it was little vessels such as the one belonging to the Prior of Tynemouth, with their prosaic cargoes of coal, salt, hides, and ill-made pottery that enabled medieval communities to break out of the limitations of their enforced self-sufficiency.

While such coastal voyaging and short sea-passages were limited to the summer months and there were lengthy delays while waiting for fair winds and the repair of broken tackle, by 1369 trade in coal had so grown that the City of London appointed four coal-meters. This was the title given to officials who supervised the unloading of cargoes, collected duties imposed by the City and Crown, and protected buyers from short measure at Billingsgate, Queenhithe, and Dowgate, which were the only official medieval landing places for bulk cargoes within the City boundaries. Coal was measured out by a vat of standard size to prevent the giving of short measure. While the trade was not yet extensive enough to require a market and guild of its own, it was officially placed, as medieval practice insisted, in the hands of the Guild of Woodmongers. The coal-meters, the first of a numerous and unloved race of officials did not, of course, exist merely to protect the customer from the wiles of the shipmaster bent upon giving short measure. They provided the basis for the collection of local dues at the port of discharge. It was not long before the growing development of the trade and the demand for coal by numerous buoyant industries, ensured that it drew the attention of the tax-collector. A more unusual tribute was exacted by the Scots in 1320 when they demanded, and received, a ransom to be paid in sea-coal for the exchange of a

At Whitby, in the 'Haven under the Hill', there remained an atmosphere of wooden ships and the trades that served them long after it had vanished from the Coal Ports. It is noticeable how the traditional flat-floored brigs sit squarely on the dried-out harbour while the sharper built brigantine (right) cants over.
Sutcliffe Gallery

King's agent. The Mayor and Bailiffs of Newcastle provided 1,000 chaldrons and the prisoner was freed.

The quickening pace of the trade was also evident from increasing activity along the Tyne. Pits were opened at Caundon in 1350, Lumley, Cockpen, and Rainton on the Wear in 1375 (in the latter case in connection with the production of salt), while the coal-burgers of Gateshead secured the intervention of the King to prevent their commercial domination by the lords of coal in Newcastle. It had been noted as early as a hundred years previously that the approaches to Newcastle were dangerous because of unfenced excavations for fuel. Industrial man had indeed arrived in the North East.

The export of coal to the Continent began at this period, particularly to the Northern and Atlantic coast of France, which were as isolated from fossil fuel as was South East England. The area around Liege, on the wide, gently flowing Meuse, was one of the few mining areas on the Continent whose production could compare with Newcastle at this date. Coal from Liege, with the Meuse and Maas to serve as its Tyne, competed with fuel from Durham at Calais and the Channel ports further west. French vessels bringing corn from Pontoise, far up the river Seine, returned with coal as early as 1325. The French vessels must have found the Tyne a difficult river to enter and navigate. The bar at its entrance, the Herd Sands, and the Bellhues Rock to the north, had to be negotiated and once inside they found the river itself was shallow, beset with sandbanks and in other ways ill-suited for shipping. Newcastle's ancient stone bridge effectively prevented further navigation upstream by seagoing vessels and remained a barrier for many more centuries until swept away by a spectacular flood in 1771. These problems were at least partly overcome by the use on the river of shallow draught lighters, known as keels. They and the keelmen deserve a chapter to themselves.

2 Keels and Keelmen

"Beer for the Keelmen . . . . . £1.4.0."
Collier's Account Book 1734.

THE keels brought the coal from up-river collieries, down the water highway of the Tyne, to the waiting ships below Newcastle's stone bridge and to Shields, at the mouth of the river, where the colliers were moored. The keelmen and their craft until their extinction in the mid-nineteenth century were a distinctive, if turbulent, feature of the river life for over five hundred years and even something of a legend in their own time. Like most water-borne workers they developed as a cohesive group, resentful of interference, jealous of their privileges and when ashore living in their own distinctive communities. Their life was governed by the tides, not regulated by clocks, and if they seemed to their contemporaries over-aggressive and much given to cards and hard liquor, they were ever at the behest of two demanding masters.

First, there were the 'fitters' who dealt directly with the pit owners, hired the keelmen on a yearly bond, paid them and owned the keels. Officially they were members of the Company of Hostmen. This medieval title, given to similar groups of traders in other ports, was said to derive from their original purpose of acting as host and guarantor to visiting merchants negotiating the purchase of coal. In exchange for a monopoly in handling, they paid the Crown a varying amount, which was re-negotiated throughout the centuries, for each Newcastle chaldron handled through their tightly-knit organisation. Between 1536-1635 a chaldron was 42 hundredweights. They were essentially middlemen, as little loved as most of their kind and beset with archaic documentation, reminding us that this is not exclusively a twentieth century evil. They were inclined to show little sympathy for the keelman's probelms of grounding on neap tides which scarcely ran, and spring tides which left a keel high and dry upon one of the Tyne's numerous sands.

If by any chance a keelman sought for commiseration in the misery of his lot he would have gained a sympathetic hearing from the earliest recorded watermen of the Tyne. These were the *Bararri Tigrisienses,* expressly brought to the river in the fourth century A.D. and stationed at the Roman fort at Blyth. Marched from the extreme limit of the Roman world to its northern frontier, in accordance with normal Imperial policy, they were employed lightering military stores upon the Tyne to the Wall forts, employing the skills acquired on the distant river Tigris in Mesopotamia. There is little doubt they counted coal amongst their varied cargoes, but one is left to speculate upon the nature of the vessels they worked through the shallows and runs, to Chesterholm, the Roman *Cilivnum.* Were they built in accordance with local tradition, implying at that date an elaborated dug-out canoe, or did this northern river carry the large coracles like the *guffah,* that boatmen of the Tigris constructed and paddled upon the great waterways of Mesopotamia? The *guffah* had been recorded at the time of Sennacherib and they are still in use today. Auxiliaries of the legions were expected to exploit the advantages of their special skills and the *guffah* is certainly an exceptionally light, easily repaired craft which would have been advantageous when portages had to be made over shallows.

At the end of the eighteenth century keelman's outward journey from the staithes above Newcastle's ancient bridge at which the keels loaded, lay the anchored colliers. There awaited a captain, pacing the deck and cursing the wasting of a good northerly wind and a high-tide over Tyne Bar while beneath his feet the hold of his ship was still half empty. The keelmen could expect few condolences upon the hazards and difficulties of river navigation from the captain and mate of a collier who had recently nursed a leaking vessel through a North Sea gale for days on end.

The keelman's employment tended to be seasonal, particularly in earlier centuries since demand was less pressing in winter as shipments of coal fell off, or ceased altogether and when the unhandy colliers of the period were lying wind-bound. At the best of times the interruption of coal deliveries from the mines, whether by the design of pit-owners seeking to create a favourable market or as a result of natural calamities, conspired to make employment intermittent. Indeed, the expansion in the number of keelmen to as many as 1,600 by the early eighteenth century, was made possible by immigrants from Scotland who returned home in the winter when the trade was insufficient to support such a large labour force in regular employment.

They were considered a trained reserve, like the crews of the colliers, to be drawn upon in an emergency. During the French wars of the late eighteenth century, the Naval Impress Service, the press-gangs' official title, eyed them with an unashamed smacking of lips. Keel crews were officially protected from service but in a time of crisis, as occurred in 1803, the Tyne Regulating Officer, responsible for impressment, brought the whole coal-trade to a stop by capturing 53 keelmen whilst the remainder struck until their mates were released. He reported to his Lordships at the Admiralty: "There are a vast body of fine men in the keels who are protected by their Lordships. Capt. Skene and myself determined to take as many of the young ones as we could this morning, leaving all Skippers. I beg to observe, it is the general opinion here, that nothing will be done on the River, until the Keelmen are again protected, in the mean time we will get all we can". At this point the Mayor of Newcastle sought a solution to the thankless task of trying to reconcile the demands of the Admiralty for recruitment and the natural disinclination of the seafaring community to submit themselves to the grim prospect of life in the Navy. He hopefully suggested in 1803 that 'volunteers' to the number of 80, that is one in ten of the keelmen, should serve. This indicates how the number of keels and keelmen had dropped over a century. Eventually patriotism was complied with and the realities of the situation were faced by instituting a levy from the keelmen and coal-owners to provide bounties for the potential, if reluctant, Jack Tars and trade re-started. This was a considerable cost as the value of bounties rose rapidly; for not only the Navy, but the Army as well, scoured the area for recruits.

Benwell Staithe on the Tyne 1844, by T. H. Hair. This was one of the last staithes to use keels which may be seen alongside the loading spout. *National Coal Board*

A similar arrangement was negotiated with the other Tyneside workers such as the wherry-men, and coal-trimmers, and other river-workers, but the Admiralty drove a harder bargain with the keelmen on the Wear. Not only did they have to provide a quota of substitutes, two landsmen to count as one prime seaman, but those who newly joined the trade and those who were keel crewmen and subsequently reached the age of eighteen, when they were liable for impressment, had to contribute to the levy. The cost of an individual's substitute, if one could be found and he must not be accepting service in the Navy to escape imprisonment, rose from about £25 to as much as £60 by the end of the wars. This was far from a light burden upon a group of workers whose weekly wage was counted in shillings.

The keels themselves were as full of character as their crews, and peculiar to the Tyne, Wear and Blyth. Parallels to them could have been found, until a generation ago, working upon every navigable river in Britain. Trows on the Severn and Avon, Humber sloops, wherries of the Norfolk rivers, were all survivors of a time when rivers played a vital part in the transportation system of our nation. Some, like the Thames spritsail barge, evolved into sea-going craft; others, like the Rother barges in Sussex and the barges on the Teign, never grew beyond the limitations of river work. They, like the Tyne keels, hardly developed beyond a simple medieval form; in the case of the keel markedly similar to a vessel found in the mud when the moat of Kalmar Castle in Sweden was drained in 1932-34, and dating from the middle of the thirteenth century.

As in the boat found at Kalmar, whose dimensions are almost identical, a pointed stern enabled the keels to be easily constructed and assisted their handling under oars; of relatively shallow draught, drawing when fully loaded only about four feet, six inches, they were forty feet overall and extremely beamy; no less than nineteen feet amidships. The unique feature of the Tyne keel was not in her construction or antique form, however, but her identification with an official measure of coal nationally accepted and determined as early as 1266. A keel load at this time was 20 chaldrons, a measure of capacity, and approximately 17 tons. This measurement lasted until 1497 when there began a long period of bewildering variations until 1635, when it stabilised at eight Newcastle chaldron totalling 21.1 tons or after 1685 rather less than 16 London chaldrons. The Newcastle chaldron derived from a horse drawn wagon or wain loading 17 cwt. All keels were measured, in theory at least, by the King's Commissioners, who filled a newly built craft with a previously determined amount of coal, and a load-line was then marked "with a stock nailmark" upon the hull; the practice on the North East coast of selling coal by the keel-load was usual and continued until the nineteenth century. Merchant vessels, particularly on the East Coast, were stated in their documents to load so many keels of coal and were advertised with this information prominently displayed. The Newcastle chaldron was used as a standard

The model of a keel in the Black Gate Museum, Newcastle upon Tyne, showing the open hold and the boards used to secure the cargo heaped amidship.

Photograph from Society of Antiquaries, Newcastle

measure for the purposes of payment of dues at Lynn, Great Yarmouth, Southampton, and all along the East and South coasts. London dealt with a chaldron of 28 cwt, which had risen from 21 cwts in 1530. These changes resulted from technical innovation, which brought about the use of larger containers both above and below ground for handling newly mined coal, and efforts to evade taxes.

Before the exploitation of the riverside coal-seams the merchants of Newcastle looked to ship, lead, and hides as a primary source of their wealth, and the earliest mention of a keel is in 1330 when one was stated to be loading wool. But only a little later in the fourteenth century ecclesiastical mine owners of "Tinmouth" and Durham employed them for coal cargoes and soon we hear of the keelmen as a 'craft and fraternity' in the medieval pattern.

As the production of riverside coal-workings and pits within easy hauling distance of the Tyne increased, so did the numbers of the keelmen. The number of 1,600 achieved by the early eighteenth century probably indicates the existence of some 500 keels as the crew, often working under a skipper, consisted by two of three 'shovellers' and a boy apprentice. This date marks the peak of their use. The apprentice, usually related to one of the crew, was known as a 'pee-dee', a title which has defeated etymologists who have attempted to discover its origin.

The keelmen began their work by loading coal into the open hold of the keel from the riverside 'spouts' as the wooden chutes were called and down which the coal tumbled, at Prudhoe, Elswick, and the Team Gut where the Derwent and Tyne meet, to name only a few. The extraordinary beam of the keel was essential to maintain its stability, for the hold floor, supported on massive beams, was only two feet below the gunnel. This facilitated 'casting' as loading the colliers was called; the difference in height between the deck of the ship and the deck of the keel being then as little as possible. The freight was left piled high amidships for the river journey; no hatch covers were fitted, and lengths of board were used to retain the sides of the heaped coal. Latterly keels were constructed with one of the side-decks narrower than the others. This asymmetrical plan enabled the keel to be moored alongside a collier with the minimum of distance between the heaped coal and the ship's deck, although it could scarcely have improved the lighter's seaworthiness.

Dropping down on the ebbtide from the up-river pits, the keel was propelled by a pair of oars and steered over the stern by a 'swape' fully 25 feet in length. Ironshod poles called 'puys', worked by the crew from the waterways on either side of the hold, assisted the vessel over the shallows, and a single grimy square-sail was set when the wind favoured. In later years a spritsail and staysail were carried, together with a rudder, which improved the keel's performance when working under canvas, but the tide still provided the main motive force.

Once moored alongside the anchored collier, the gruelling work of casting up the coal by shovel commenced, one keel working on each side of the ship and continuing, if necessary, after nightfall. As darkness fell, the blackened, sweating shovellers laboured by the glare of coal-fired braziers flaring and dying as the wind came in from the sea. Later a removable section of the collier's bulwarks, opposite the main-hatch, called a gate, enabled the work to be made somewhat easier, and the biggest vessels must have had ports in their topsides for this purpose, which would nevertheless have had only a limited advantage. Many colliers built expressly for the trade during the eighteenth century, like the *Brotherly Love* which survived so long, had no solid bulwarks, only a rail. As the difference in height between an unloaded collier's deck and a loaded keel's gunnel could have been as much as eight feet the work of casting must have been arduous in the extreme. Having completed discharging their own cargo, the keelman was expected to assist the crew of another keel of the same owner.

Small wonder that one of the demands made in the keelmen's strike of 1819 was for an extra shilling per keel to be paid for every foot of the side of the ship above five feet. This must have been a longstanding and mounting grievance as the tendency for vessels in the coal trade to become bigger and inevitably higher as time went on, made loading progressively more difficult. The average tonnage of coasting vessels had doubled between 1700 and 1800 and the Tonnage Laws, by emphasising beam and length, tended to produce narrow deep vessels with high topsides.

Although keelmen had "to furnish and supply their own necessary and proper "Working Geer" it would seem that a keelman's wages were somewhat in excess of the contemporary agricultural labourer, and were augmented, as most earnings were in the eighteenth century, by official and unofficial perquisites. This extended to fuel and beer, but a macabre sideline is indicated by an entry in the accounts of Newcastle Corporation in 1837-8 - - "To Keelmen, for taking dead bodies out of the river £18.13.0d."

As long as the mines above Newcastle's low-arched bridge produced the bulk of the sea-borne coal all was well with the keelman's traditional way of life. But the exploitation of pits lower down the Tyne began about 1800 with the steady introduction of steam pumping and the extension of tramways, and this enabled coal produced several miles from the river to be brought to the staithes. It could then be loaded directly into the colliers lying in deeper reaches of the river by drops, an innovation which forced the keelmen upon the defensive in an attempt to preserve their long-established trade. Traditionally a belligerent group, in 1671 Gateshead burgesses had "Paide for powder and match when the keelmen mutinyed 2s", they struck. When defeated in industrial action they entered upon long and expensive litigation in a forlorn attempt to defeat an on-going technology.

Strikes continued intermittently, and during the most extended of these, in 1822, the keel owners installed one of William Hedley's pioneer steam locomotives in a keel. This was equipped with primitive paddle wheels, geared to the driving axle, and steam towage on the Tyne received a new impetus. The arrangement not only successfully propelled the keel on which it was mounted but towed a string of loaded ones astern, much to the chagrin of the strikers. In less than a decade, by 1830, Marshall's shipyard at South Shields had begun to specialise in the construction of steam tugs, not only for the Tyne but for ports far afield. The story of the Tyne's tugs is further dealt with in Chapter 9.

In 1799 there were 320 keels plying the Tyne compared with their highwater mark of 500, and the downward trend continued, although the Wear still employed 520. The days when 55 keels had loaded at a single up-river staithe were gone. The "Coal Drop and Spout at Wallsend", from *Views of the Collieries in Northumberland and Durham* by T.H. Hair, dated 1844, shows how sophisticated

A drawing by T. H. Hair of the drops and spouts at Wallsend on the Tyne where the house-coals were loaded. These drops were developed after 1800 and lowered the waggons of coals to the ship's deck before being emptied, thereby causing less breakage of coal. They caused redundancy amongst the keelmen but the final loading had to be completed with their assistance. *National Coal Board*

the loading arrangements which superseded the keels were to become on the lower reaches of the river. However, one may speculate whether the larger number of keels surviving on the Wear was due to the acceptance of the innovation of loading coal into iron tubs, each holding a chaldron, which were then conveyed down river in keels. These were then lifted by either a shore or floating crane into the hold of the ship and emptied by labourers. An early form of containerisation!

Despite the innovations due to industrial development, keelmen and their craft still had their uses. There was a limit to the loading that could be undertaken by a collier lying at a staithe however ingenious the method. The staithes projected out into the river and so were necessarily limited in length to avoid obstructing navigation; they terminated in relatively shallow water, and to prevent grounding as the vessel's draught deepened, a collier would warp out from the staithe into the stream and complete loading her freight from the keels in the traditional manner. The term 'staithe' first derived from the Old English 'stoep' meaning a simple landing place, and had in turn come from the Old Norse. It is used with this meaning as far south as Norfolk but is then replaced by the Saxon term 'hythe', with a similar implication.

But with the introduction of steam-towage and the energetic improvement of the river after 1851, the keel and the keelmen were finally doomed. Something bigger than the traditional 21-ton keel, evolved in the days of very restricted production of coal, was required and this brought about the development of the wherry, capable of carrying the cargo of several keels, deeper, but basically of the same design. These were equipped with living accommodation, known as the 'huddock' and, like the keels, still occasionally set sails. The keels were always carvel built, strongly constructed with grown oak timbers and planked with oak to withstand the rough treatment they undoubtedly received. In the best examples a man's fist was scarcely expected to fit between the ribs, so closely were they spaced. One peculiar feature of the later wherries was that their builders reverted to the use of clinker planking, showing how deep-rooted was this tradition in the North East and a reversal of the usual pattern of development in river craft. Perhaps if offered greater economy in construction. In 1973 it was announced that the Maritime Trust is preserving one of these on the Tyne.

By the mid-nineteenth century only four of the score of collieries in County Durham and four of the 23 in Northumberland, amongst them Liddell's Main, Shipcote Main, South Moor, and Gorsefield were still moving coal by keel. Finally, the removal of the old stone bridge at Newcastle in 1876 and its replacement by a swing bridge, together with the continued dredging of the river, enabled vessels to load full cargoes at the staithes. With their purpose gone, many keels found honourable retirement employed as landing-stages for the Tyne passenger steamers and a few found use with riverside industries as towed lighters. A search conducted thirty years ago failed to reveal a single survivor; like the keelmen they had been superseded by a better ordered, but less colourful, way of life. The rapid industrial growth of Tyneside during the second half of the nineteenth century no doubt absorbed the crews of the keels without much difficulty, as their calling and craft became redundant. One wonders, nevertheless, whether they did not look back with regret upon their lost heritage as keelmen, incarcerated as they now were in mine, brick-works, foundry, or the iron fo'c'sle of a screw collier. Plodding at dawn along cobbled streets to the pit-head they must have recalled past mornings when the Tyne glittered at the foot of the 'chare', as the alleyways leading to the Quayside at Newcastle are known, and the tide ran seaward.

One substantial relic of their heyday has not been lost. The keelman's hospital in the City Road, Newcastle, still stands, quietly dignified, with a distinction about its pilastered tower and sash windows speaking of a Foundation in the days of Queen Anne. With high and laudable ambition in 1699 the keelmen arranged for a deduction to be made at the rate of fourpence per keel, per tide, (the journey to Shields and back), from their wages. With these monies the charitable hospital for their sick and aged was built by 1701. The Scottish keelmen, being migrant labour, did not qualify by residence for the meagre parochial relief available in the seventeenth century and the Foundation was of particular value to them. Alas, dissension and distrust between employers and employees ended the scheme with mutual recrimination in 1712. However, the admirable work of the Charity survived to be taken over by Newcastle Corporation in 1872, a fitting reminder of the Tyne keelman's loyalty to his fellows.

Hev ye seen owt o' maa bonny lad,
And are ye sure he's weel, oh?
He's gyen ower land wiv his stick in his hand,
He's gyen te moor the keel, O!

Yes, aa'v seen yor bonny lad
Upon the sea aa spied him,
His grave is green but not wi' grass
And thou'lt never lie a-side him.

3 The Emergence of the Collier

THE impetus which the production and radet in sea-coals had gained during the thirteenth century seems to have spent itself so that even three hundred years later there is little evidence of substantial growth. Of the 15,000 tons produced annually, approximately half found its way down the sea highway to London and the out-ports. Although London's Billingsgate Quay today is synonymous with fish it was there that a coal-meter was appointed in 1444. Coal cargoes unloaded there, under his or his assistant's eye, were conveniently placed for the City and it was to be many years before the quay was monopolised by fish traders. A contemporary witness noted a lively scene with "Frequent Quarrels amongst Coal-heavers, Porters crossing with Burdens, Fish-women, Wheel-Barrows, Quack-Doctors and Barbers".

Less conveniently, but at greater leisure and without the restrictions of guilds, coal cargoes from the North East ports were landed on a hundred open beaches between Tees Mouth and Hastings. The smaller ports, some like Dunwich on the Suffolk coast lost to the sea, others like Sandwich in Kent forsaken by it, collected coals from the holds of the early colliers; the loaded tumbrils creaking and clattering away to the local lime-burners kiln or blacksmith's shop. Coal was gradually becoming a growing necessity and not merely an intermittent need.

Newcastle and its immediate neighbourhood whence the coal-ships came, were the only area in the whole of Britain where coal dominated the scene. The ancient town jealously guarded its position of holding a near monopoly in 'sea-cole' and deliberately inhibited the potential production and trade of comparable places on the Wear and the Northumberland coast. But the growth of salt manufacture had begun at South Shields, at the mouth of the Tyne, and further north at Cullercoats, Hartley and Blyth. This industry used 'small coals', which were looked upon with disdain by the Guild of Hostmen at Newcastle. Nevertheless they were acceptable as a source of heat beneath the lead or iron pans in which sea-water was evaporated and the resulting salt collected for shipment to Hull and the fishing port of Yarmouth.

It was the demands of this primitive industry, so essential to medieval domestic economy for the preservation of foodstuffs, that gave a stimulus to mining at Wearmouth, outside the Tyneside area. In the mid-fifteenth century the monks of Tynemouth secured a Charter to legalise their trade in coals and white salt and the manufacture of white salt must have been one of the earliest associative industries of the Great Northern coal field. The name 'white salt' distinguished it from the inferior 'grey salt' of the East Coast, produced at least partly by solar evaporation in marsh-land salt pans and inevitably containing a fair admixture of marsh-land mud. Salt cargoes were popular with ship owners. Some found its way into the bilge of the vessel where it was said to act as an excellent preservative for the timbers.

Ever since early medieval times, when the progressive clearing of the primeval forest had taken place, the woodlands had been intelligently cropped in order to conserve this vital raw material. The 'Great Re-building', which took place nationally in the second half of the sixteenth century further exacerbated the shortage for it has been calculated that for every yeoman's house built, two acres of mature woodland had to be sacrificed to the axe. As a result, a rapid change in the status of coal took place; from an undesirable fuel it became a vital element in the nation's economy. The prohibition of its export was frequently discussed. But such a royal order as that made in 1546 indicates the scale to which the sea-borne trade had grown, in that it was commanded that 3,000 chaldrons, some 100 ship-loads, should be dispatched to Boulogne with all haste. However, the requirements of State were permitted a priority not allowed to civil traders. Only a little later the export of coal out of the realm was firmly prohibited. Coal, the order stated, was becoming a frequent ballast cargo "and resulting in a most exorbitant dearth and scantiness of fewell". When one learns that it was being bought for 2s.2d. per chaldron on the Tyne and sold in France for £4.6s.8d. one can appreciate the eagerness of these Tudor merchants to export. The official view that it was a ballast cargo would indicate a certain slowness of the administrative mind to recognise that coal might be as remunerative a freight as wool or wine. This export is an early pointer to the growth of the Newcastle coal trade and how it had become as important to France as it was to London. The earliest pictorial evidence I have discovered of this export trade is a French one; a medieval psalter dated 1341 showing barges alongside a Paris *quai* unloading sacks of coal. However much the odour of apple-logs or the scent of glowing pine boughs burning upon the hearth might be preferred to the stink of coal-smoke, sheer necessity brought victory to the black diamonds.

Competing with the demands of the growing numbers of domestic hearths and the industrial furnaces of the metropolis, was the ancient Wealden iron industry of Kent and Sussex, geographically near enough to make wood-fuel increasingly scarce for Londoners. All three groups of consumers brought down upon themselves the censure of the Tudor shipwrights, struggling to meet the royal demands for increased naval construction, and who were appalled at the wanton destruction of so much irreplaceable oak and elm. The fuel-using industries, such as dyeing, brewing, brick and glass manufacture, were beginning to emerge as well organised activities. They sought coal in a steady supply. Although the Company of Brewers, doubtless then, as now, sensitive of their public image, loyally undertook to use only wood in the vicinity of Westminster Palace, to avoid "nuisance to her majesty Elizabeth", they all needed coal for their industries and upon an ever-growing scale to keep pace with the increasing demands of their trade. As the brick chimneys, each more richly decorated than its neighbour, became an accepted feature of the fine new Tudor mansions, London made one last gesture of repugnance at this new fuel, dug from the depths of the earth. It grew only westward, up-wind, so that the prevailing south-west wind took the sulphurous smoke away from the most recently developed surburban belt. 'South-of-the-River' was doomed to a smutty and therefore unfashionable semi-industrial image, from which it has never quite escaped.

By the second half of the sixteenth century the quantity of coal shipped over the bar of the river Tyne had risen from 35,000 to 160,000 tons. Newcastle merchants and coal masters were enlarging the area from which coal was mined. They had done well from the desportation of the church land by Henry VIII, indirectly securing leases of collieries on former ecclesiastical land at Stella and Ravensworth. This was a boom time for the industry and it was centred upon the Tyne, almost entirely above Tyne bridge, with the burgesses of Newcastle jealously guarding their virtual monopoly in mining. This monopoly was to be further enhanced by a charter, granted by Elizabeth I in 1600, giving the guild of coal merchants almost unassailable powers. The Hostmen's Company of Newcastle, with the coal trade secure and the government of the port and river in their capacious pockets, were rightly called the Lords of Coal. Mining development had been previously restricted to the manors of Gateshead and Whickham, on the south bank of the Tyne, between the mouth of the Derwent and Tyne Bridge. Now the extraction of coal began from pits which were sunk on the north side of the river at Newburn, Denton, Elswick, and Benwell.

But the problems of transporting the mined coal from pithead to riverside staithe remained a stifling limitation; no pit was worth exploiting should it lie more than two miles from a point where its production could be loaded into the hold of a waiting keel. However, it must be emphasised that the Tudor and Stuart age was much less enamoured than our own with the alleged benefits of a constantly rising scale of production. They were too suspicious of the changes it would bring and the insecurity which the consequential social movement involved. State intervention, when it was used, sought above all to regularise and stabilise, and to ensure that if change had to come it was supervised closely by those who had proved their responsibility in the past.

As the coal-trade grew in volume and importance the absence of any specific guild to supervise its regulation in the North had been overcome by allowing the Hostmen of Newcastle to undertake the duty. The original duty of the Hostmen had been the supervision of the town's trade with visiting merchants and so the expansion of their activities to this wider field of management was a natural one. By a statute of 1529 they were given the right to insist that only at Newcastle should any ship load or unload, or goods be sold into or from any vessel on the Tyne. As ships became larger they found the undredged channel of the river impossible to navigate as far as Newcastle and Shields became an accepted anchorage by the seventeenth century. The fact that all cargoes were loaded from keels, owned by the Hostmen, enabled the guild to continue its tight grip upon trade.

North East coal producers were forced to bring their output to the town and were only permitted to sell to freemen, never direct to the ship-owners whose vessels waited at Newcastle Quayside. The Hostmen naturally found it expedient to become coal-mine owners where opportunity offered and gathered into their fold other freemen of Newcastle who already had interests in pits. As early as 1603 they took the next obvious step and a carefully organised restrictive arrangement was evolved. This limited annual production to a little under 200,000 tons, a ceiling reached in much less than a full year's working. It had the immediate effect of raising the price of coal in London and it was perhaps unfortunate for the Hostmen that their principal market was also the seat of government. The Privy Council forbade such a creation of artificial scarcity, but the order failed to have a lasting effect and a manipulation of the market continued under one guise or another for a further two centuries.

To regulate waterborne affairs the medieval system of controlling guilds was adapted rather more happily to a new and expanding economy. Henry VIII granted a charter which served to consolidate the long established guilds who had governed and ordered the shipping and seamen of the port, into one Trinity House of Newcastle. There is no doubt that the Trinity Houses of Newcastle, Hull and the Thames, the body which eventually gained national supremacy, had roots in late medieval times. The charter given to Newcastle was later to be confirmed by Elizabeth I and its activities strengthened; in the almshouse 'aged and impotent seamen' found a refuge, apprentices were taught seamanship, navigation and Godly fear in its nautical school, while pilots were examined and licenced, and the craft and mystery of seagoing furthered by its Elder Brethren. All met regularly to worship in the private chapel. The chapel and related almshouses are still to be found near the old Quayside, east of the bridge, the chapel resplendent with its Jacobean carving dating from 1634, rebuilt into a panelled hall of a century later. The Guild, Fraternity or Brotherhood of the most Glorious and Undivided Trinity of Newcastle shared the jurisdiction of the Port of Tyne with the Corporation of Newcastle and, at least in theory, used the dues produced from the light-houses and pilotage to further charitable work and to defray the expenses of administration. This was a continuously expanding income for it was a source of revenue based upon the tonnage of passing shipping. Besides controlling the lights of Tynemouth they maintained beacons as far afield as Blyth and Whitby. At least some of these had been acquired from ecclesiastical houses at the time of the Dissolution and Trinity House perpetuated a medieval pleasure in pageantry by furnishing a rowed barge to augment the annual Ascension Day water-procession upon the Tyne.

Afloat a change had come rapidly and with far-reaching effect. The appearance of the ships, large and small, which moored against Newcastle Quayside, or waited apprehensively for high water to take them over Tyne bar, showed that nothing short of a revolution had taken place in the shipwright's craft. One of the earliest references to a specific coal-trade fleet is in 1596 when the first proposal for an armed convoy of colliers was made and it was stated there were some 200 vessels then engaged in the trade. This fleet would have consisted of crayers (meaning carrier), hoys and ships, the latter probably engaged for loading a 'dirty' cargo fitted in between more lucrative voyages, further afield. All, except the smallest, would be carvel built in the new method, a style of construction originating from the Mediterranean and reaching North Western Europe via France. It was only adopted reluctantly after the older clinker method had been pushed to, and beyond, its limits in the attempts to build larger vessels.

This new style of building seagoing ships required a basic ability to project in scaled-down form the shape of the hull. This could be done either two dimensionally in drawings, or three dimensionally in a model. It needed an ability, going beyond skilled craftsmanship, to reproduce the shape so that on the keel the frames could be set up and the stern and stern posts erected at the appropriate rake. Around the frames were attached the planks which formed the skin of the vessel and the subtle form of these, particularly difficult to achieve at the stern, had to be calcualted with little margin for error. To make a watertight joint between these planks caulking was used. This consisted of forcing strands of spun oakum into the outer side of the joint with a chisel-shaped iron, struck by a wooden mallet; the distinctive sound produced was to ring from the shipyards of our harbours and rivers for as long as wooden ship-building lasted. From the seventeenth century onwards timber was measured by the load; this was the amount that could be drawn by one horse and approximated to one ton or forty cubic feet. Between one and nearly one and one half loads were needed for every ton of commercial shipping constructed. English oak remained the supreme timber for ship-building but had to be augmented to an increasing extent with oak imported from the Baltic, fir and beech. Elm was the traditional material for the keel because of its long run and water resistant qualities.

The use of the lead-line was always an important element in East Coast navigation. Leadsmen stood in the chains and called out the depth of water indicated by the run of line. *National Maritime Museum, London*

The timber frames were made up of several pieces, and built upwards from either end of a 'floor' as the middle part, laid across the keel, was called. As work proceeded the whole structure became more complex; the deck-beams locked together the tops of the frames, beams across the hold strengthened the hull transversely and the keelson lay directly above the keel, securing the floors and giving longitudinal reinforcement to the hull. Great wooden knees, sawn from naturally grown crooks, secured stem and stern post to the elm or oaken keel then others were fitted above them to make doubly sure. Down below the hold was lined with pine boards and above, over the deck beams the deck was laid, forming an integral part of the structure. The deck had as few apertures as possible; hatchways were a constant source of weakness for while a merchant wanted them large to facilitate loading and unloading, the seaman insisted that, in time of stress, the fewer points at which the hungry sea could enter the better. The same reasoning applied to the cutting of loading ports and gun-ports in the sides of the hull. They were provided with reluctance, for seamen knew that they not only weakened the hulls structurally but provided the chance for wet-rot to take hold in the airless crevices of the ship.

But a wooden ship is only as good as the fastenings which secure planks to frame, keel to apron, decking to beams, and riders to keelson. This fastening was done by long wooden pins, called treenails, exactly as had been used in the older form of construction; they were turned from oak and might be between one and three feet in length. The hole to take them was made by an auger of the appropriate size and then they were driven home with a maul, a task requiring great skill. Deck and rails were fixed, more prosiacally, by nails and iron spikes.

The building of a ship required a wealth of skills; men to draught the plans, men to sweat at the great pit-saw, men to scarf together timbers, boys to bring beer, women to strand oakum, old men to mix tallow and resin for the bottom, mast-makers to dub trees with axes, riggers to work rope with marlin-spikes aloft, and caulkers to swing mallets below. There were rope makers and block makers, sail makers and coopers, blacksmiths and carvers, all working their various crafts close by the shipyard. By early Tudor times these skills had become fully developed and were to stay, with little change, until the coming of iron and steel four centuries later.

While the change from the old style of northern clinker construction took place relatively rapidly, it was not accepted everywhere. There is evidence that at first only the topsides of vessels were built on the new system, so allowing the more difficult, underwater part of the hull to be constructed by the older, better-understood, method. At least this hybrid construction produced a hull which would lie against a quay or another vessel with the minimum of damage; the irregular surface of the clinker construction, with its overlapping planks, must have made for very uncomfortable bedfellows when lying alongside in harbours. Some of the smaller craft never changed. The little colliers of the Humber, the 'billy-boys', were built clinker-fashion until the latter part of the nineteenth century, and were said to be the largest vessels so constructed in North Western Europe.

The growth in the complexity in the sails and rigging of the coal fleet in Tudor days would have astonished the seaman of a century before, while subsequent evolution was so slow that the rigging and canvas of a sixteenth century ship could have been handled with confidence by the bosun of a tea-clipper. The crayers, and amongst this class were the smaller square-rigged vessels, carried a main-topsail as the only canvas above the lower sails. These lower sails were lashed to yards slung across the mast and furled by seamen going aloft. However, the Mediterranean had provided more than a new style of shipbuilding, it had passed on the idea of a sail which set on a yard permanently carried inside the rigging and slung at 45 degrees to the mast, in line with the keel. This triangular lateen sail was set on the after mast, by now known as the mizzen; its prime purpose was not to act as a driving sail but to balance the sail plan, assist in manoeuvring the vessel and help the helmsman in his struggle with the great tiller.

At the bluff bow, set over the beak head was a spar known as the bowsprit. This had a long ancestry and had been carried on ships for centuries. Previously, in medieval times, it served as an extension to the bow so that ropes leading down from the edges, leaches, of the large, single squaresail could be taken to it. By shortening on one of these, they were called bowlines, and allowing the other to run free, the problem of controlling a large area of recalcitrant canvas was simplified. Above all the bowsprit assisted in giving a good set to the square sail when it was in line with the keel and enabled the vessel to sail effectively with the wind abeam. While this might seem a somewhat recondite, technical point, it meant that the capacity of the vessel to proceed upon its voyage was considerably enhanced. By now, besides serving as a point to which the bowlines were led, the bowsprit carried a squaresail of its own, known as the spritsail and, despite its apparent unhandiness to modern eyes, served well enough in helping to work the ungainly hulls of the early cargo-carriers to windward. For it was in their ability to go to sea and make their way, however slowly and with whatever expenditure of muscle and patience, against the wind, that these early sixteenth century vessels had a mastery over their predecessors. But despite this advantage, the masters of these little vessels, the balingers and crayers, hoys and cromsters which served the coal trade had, like Chaucer's shipman

> . . . to reken wel his tydes,
> his stremes and his daungers hym besides,
> His herberwe and his moone, his lodemenage . . .

The degree of seamanship required to manage a Tudor vessel was demanding, but the knowledge of tides and tidal streams, of channels and swatchways, good holding ground where a vessel could lie at anchor safely, or harbours which dried out dangerously, required a lifetime's hard-won experience.

Inevitably the majestic battle-fleets and the great traders exploiting and carrying the wealth of the newly discovered continents caught the attention of the contemporary writers and artists. The humble but equally necessary shippers of timber, coal, pewter, and cloth have left behind them few visual records. Some of the rare pictorial records are found in the charts and estate maps of this period, as delightfully decorative as they are cartographically hazy. Nevertheless, the little drawings of the vessels the mariner would expect to find when sailing the seas shown in the charts, principally those of the doyen of map-makers. Lucas Wageneer of Amsterdam in the sixteenth century,and his contemporaries, form the basis upon which one can reconstruct the appearance of the ships themsleves.

Early in the seventeenth century the complaint was made of East Coast colliers that "for thrift's sake" they carried dangerously small crews. A ratio of one man for every eight burthen tons, dropping to one for every thirteen tons, probably in the hoys, was usual at Newcastle in the 1580's. This would imply a complement of as many as a score in quite a small vessel. But ineffective mechanical aids and cumbersome and clumsy gear, the blocks were sheeveless, for instance, required many hands to work them effectively. It was here that the hoys had an advantage over the square-rigged vessels. Contemporary evidence points to the fact that their easily-worked fore and aft canvas required fewer men. The proliferation of the hoys led, in fact, to their prohibition in Elizabeth's reign. On two occasions it was alleged that their economy of operation "led to a decay of Mariners and Ships by the marvellous increase in hoys, thus striking at the root of national security", so the government argued.

Some insight into the workings of the coal trade in Tudor times is provided by the evidence given before the Star Chamber by one Thomas Bagshawe, described so admirably in the style of the age as "Merchant Marryner". He deposed that he bought and carried sea-cole in vessels and hoys, of which he was part owner. The coals were purchased at Newcastle and to manage this he sent his servants and factors there and he also instructed them to buy for other shipowners, whose cargoes he financed. Like others of his kind in the wool trade to Calais and the wine trade to Bordeaux, he did his best to guard against the vicissitudes of the business by sharing in the equipping, victualling and hiring of a number of hoys. Men of his calibre, shrewd, robust, given to calculated meanness and surprisingly uncalculated generosity, were to be the backbone of the coal trade for many centuries to come.

4 The Newcastle Voyage

"The Newcastle voyage, if not the only, yet is the especial nursery and school of sea-men. For as it is the chiefest in employment of seamen, so it is the gentlest and most open to land men; they never grudging in their smallest vessels to entertain some two freshmen or learners."

"The Trades Increase" 1615.

THE mercantile expansion of the early seventeenth century rapidly had its effect upon the size of the 'sea-cole' fleet which was, at that time, said to number 400 vessels, one half of which supplied London and the remainder the rest of the South and East Coast ports. Very tentatively one might estimate this was as much as one-sixth of the seagoing merchant vessels of Britain. A contemporary writer draws a clear picture of the Tyne during the reign of James I. "Besides our own ships hither, even to the mines mouth come all our neighbouring nations with their ships continually. The French sail hither in whole fleets of 50 sail together, serving all the ports of Picardie, Normandie, Bretagne, etc. even as far as Rochelle and Bordeaux; and the ships of Bremen, Embden, Holland and Zealand, supply Flanders etc. with our coals". In 1616 for every one English vessel leaving Newcastle there were seven foreigners sailing out over the Tyne bar, despite the very high Customs imposed upon exported coal by both the port and the crown. But this motley company of merchantmen could, all too easily, be discomforted by sea warfare, both civil and international, while as we have seen less spectacularly delay and disorganisation were engineered by the combination of pit-owning Hostmen seeking to regulate quantities and prices in their own interest.

Sometimes religious enthusiasm could have unfortunate effects. The Mayor of Newcastle wrote in 1639 that "the Ipswich Puritans have so wrought with the shipmen that for six weeks I did not load one chaldron of coals so that my staiths are so full that they are like to fire" and he concludes, somewhat lamely, "I had one fire last year", although whether this was the result of religious heat or spontaneous combustion is not enlarged upon.

The rusty machinery of distribution was still of a basically medieval construction and if the Guild of Hostmen connived along the banks of the Tyne, they found eager imitators on Thameside where the coal trade was still under the thumb of the Woodmongers' Guild. Their water-borne market in fuel on the river, with its transhipment from seagoing craft to river lighters caused much inconvenience to shipping and passengers. Finally in 1607 it was expressly forbidden to sell coals upon the Thames in any other vessels than those in which they were first brought within the port. The seventeenth century Thames tideway was a crowded and colourful river. It served as a highway for the traveller who, aboard a rowing wherry, attempted to avoid the filth and footpads of London streets as he made his way from one riverside stair to another. The absence of sufficient quays where cargoes could be loaded or unloaded under official eyes meant that the passenger wherries worked their way between tiers of vessels which lay waiting impatiently for the opportunity to discharge their freights. It made a fine sight for the visitor but the enforced idleness must have infuriated the ship-masters.

Besides regulating the methods by which coals were sold in London, King James augmented his revenue by extending the tax upon exported coals to the rate of 3s.4d. per chaldron, should they be carried in foreign ships, but 1s.8d. if exported in English vessels, partly, at least, to give protection to native shipowners and ship-builders. His son Charles I placed a further imposition and it was said "that the farm of sea coals is the bravest the king has".

The ever impecunious Stuarts made full use of the coal-trade to raise revenue and a toll of one shilling a chaldron was granted by Charles II in 1677 to the Duke of Richmond on all coals shipped from Newcastle. It was known as the Richmond shilling and was a particular source of complaint by those engaged in the trade until finally, in 1799, it was purchased from the Duke of Richmond by Parliament. The accession of William, however, continued the tradition and a new duty, introduced in 1697, imposed a farthing per ton on all coal shipped from the North East. At the same time the grade of Customs Officer known as 'coal meter' was regularised. He was employed in measuring and weighing coal, culm (anthracite) cinders (coke) and certifying the types and quantities for the imposition of Customs' duty.

Despite the efforts of Newcastle to stifle the trade in coals from other ports which were in a position to exploit the Great Northern coal field, their share in the trade had gradually grown. Blyth had long maintained a salt-producing industry, dependent upon locally mined fuel and the last salt pans near the Keel Dock were not demolished until 1876. But the first attempts of London adventurers, amongst them Peter Delaval, to mine coal for shipment were unsuccessful. These early pits were sunk, near the river Blyth, at Cowpen and Bebside. However, the river was shallow, more suitable for salmon than commerce, and although keels were used from 1598 to assist in transferring cargoes to deep-draughted seagoing vessels, as they were on the Tyne, even this did not suffice. The final keel-loads had to be taken aboard outside the harbour entrance, a process known as 'keeling off'. Some keels actually made short coastwise voyages southward to the Tyne and Shields. This hazardous undertaking, when the open keels were at the mercy of a sudden deterioration of the weather and the ships were vulnerable to dragging on to a lee shore, lasted until 1869 when improvements to the channel made it possible for fully loaded vessels to leave the river without risk of grounding. Blyth, then, remained a straggle of fishermen's cottages and relied principally upon its smoky salt-pans for profit during the centuries with which we are dealing at present.

Unlike Blyth to the north, the River Wear lying south of the Tyne presented a more likely source of competition to the Lords of Coal in Newcastle, and during the Civil War its chance came. The complex alignment of communities and class during the war between Royalists and Parliamentarians almost defies analysis. Port communities tended to support the Commonwealth, but Newcastle, perhaps because of its remoteness from the capital, and their hostmen who owed their position to Royal Charter, opted for the King. Whichever party commanded the seas would maintain control of the North East coal trade, and this was eventually in the hands of the Parliamentary fleet. In 1642 an order was issued by Parliament prohibiting ships bringing coal and salt from Newcastle, Sunderland, and Blyth. As a result, the price of coal rocketed. Urged on by petitions from the out-ports foretelling outbreaks of disorder amongst seamen, porters and the poor if action were not taken, the compromise of freeing the ports of Sunderland, at the mouth of the Wear, and Blyth, from blockade was agreed to. But it was maintained against royal Newcastle as long as its burgesses held out against the besieging Scots. The merchants and coal fitters of the city must have felt particularly bitter when the little upstart harbour of Blyth provided facilities for the landing of supplies and men to support the Scots' army investing their ancient town. The stimulus which this enforced restriction upon the principal producers gave to mining and shipment outside the area under the immediate influence of the Hostmen of Newcastle enabled Sunderland to emerge as an effective competitor when peace came. In 1660, by which time trade had again been re-established, while there was little perceptible growth in the output of the Newcastle, Sunderland had increased production fifty times, but still only managed to export less than one-third of Tyneside coal.

But a change had been wrought in the composition of the coal fleet which was, although peaceful, as far reaching as any produced by the hostilities of the Civil War. The opening years of the seventeenth century saw an expansion of Dutch maritime enterprise which gathered momentum as the age progressed. It produced its own characteristic vessel, the *fluitschip,* or as it was known to the British seaman, the fly-boat or flute. Characterised by its large hold, carrying between 100 and 500 tons, they were inexpensive to construct, and had low operating costs. All vessels are a compromise: a compromise to obtain an economic proposition from all points of view, the owner's, the seaman's, the shipper. The flute was no exception. If cargo-capacity is sought above all other considerations, speed and sea-kindliness are lost; if rigging and sails are complex enough to enable every condition and combination of wind and weather to be taken advantage of, then mast and spars become inordinately expensive. Should the sail-plan be too pinch penny in quality or quantity the ship's performance will suffer. Crews must be numerous enough to handle the vessel in a seamanlike way, but not so many as to eat, literally, all the profit from a voyage, for while they were paid by the voyage, food and drink was traditionally provided by the owner. The construction of the vessel's hull should be such as to allow repairs to be simply effected and use made of easily available material, while size and draught must be such that a paying cargo could be quickly found and its collection should not delay a ship outside, or in an inadequate harbour waiting for high tides. For her century the *fluitschip* was a better calculated compromise than any of her British equivalents.

We, in the twentieth century, can sympathise with our forefathers' lament over this Dutch supremacy brought about by the flute in the Narrow Seas and Baltic for it has been repeated with almost equal success in our own time. As an Act of Parliament, dated 1685, stated, deploring the decay of native shipbuilding, it was due to . . . . . "the freedom which foreign ships and vessels, bought and brought into this kingdom, have enjoyed in the coal and other inland trade equal to that of English built ships". The flute was certainly a more economical proposition than any of its contemporaries, French or British. One may still detect the influence of the Dutch flute in the hull-form of the collier-bark *Liberty and Property,* built in 1752, if one judges from her builder's model. Yet this Whitby-built bark was launched fully a century after the heyday of the *fluitschip.*

The flute's rounded stern was simple to construct and sea-kindly, the floors of the hull were flat to enable the vessel to ground without damage and to assist in unloading on the shore, while the hull made no attempt to affect a belligerent appearance or elaborate decoration. The modest sail plan was similarly unpretentious. The Hollanders relied upon prompt sailing and delivery, upon cleanliness and sound gear aloft to maintain their reputation. Ashore they were supported by a sophisticated system of insurance and financing. In all fairness to our shipwrights and merchant vessels, one must quote a contemporary writer who said "The

A French engraving of Dutch fluits. The different scale on which the same hull-form and rig could be reproduced is well illustrated. *Author*

English ships will double a lee-shore when the others will be forced in foul weather upon the rocks; the cables and anchors of the English will hold when the others will be forced to come home and break. The English will be able to put out sail at sea when the others will take in sail. The English go deep in water which makes them wholesome in the sea and carry the merchants' goods with little loss". This paean of patriotic praise had a basis in fact, no doubt. However, such was the organisation of Dutch shipyards, their ancillary trades, and the superiority of their equipment, they employed saw-mills driven by windmills, one of which survives to this day near Warmond, that building costs were low. As well as this there was the easy geographical closeness to a wide range of raw materials from which hulls, masts, and sails were produced. A Dutch vessel could be launched, ready for trading, at three-fifths the cost of its British counterpart. Finally, and perhaps decisively, when financing his enterprises the 'merchant maryner' of Hoorne would borrow money at a far easier rate of interest than his equivalent at Whitby.

But perhaps the heart of the matter was touched upon in the remark of a seventeenth century Englishman when he said of Holland that it was "such a spott, as if God had reserved it a place onely to digg turf out of". The Dutchman turned naturally to the sea as a place where riches could be exploited, as a highway upon which he could trade, as a place upon which he could pleasurably exercise himself and not least as a source of artistic inspiration. The Englishman was content, at least during the seventeenth century, to find all this upon the land or highly speculative ventures on far distant seas.

So, at the time of Cromwell, traders of the North Sea had the distinctive tumble-home, round sterns and tall topsails of Holland afloat. Despite the Protector's attempts to change the tide of events by his Acts of Navigation in 1660, the foreign-built content of the British mercantile fleet grew, albeit surreptitiously. The coal trade, not of great profit to courtly private adventurers, but "breeder of seamen and nationally gainful" was, in many cases, carried in Dutch vessels. Prizes of war they may have been, or naturalised in order to comply with the New Navigation Act, which was in many cases only a paper change of ownership, or perhaps ships which had passed into British hands before the laws which discouraged purchase abroad had been passed. But prizes taken at sea numbered hundreds as this was the main source of the foreign built, cheaply run tonnage.

The coal trade suffered the disadvantage of being a one-way traffic from the staithes of the Tyne, to the quays and hythes of the Thames estuary. Only the most economically managed ships could survive in it. Of the 400,000 tons of coal and 6,022 wey of salt that were shipped in 1634 and of which some three-quarters went to London, most were carried on vessels which were hard put to find a return cargo. However, Newcastle was almost unique amongst the ports of Britain in that it exchanged an industrial product for agricultural imports. These came from little ports like Wells, today only to be entered at spring tides by the infrequent trader, or Blakeney, now only beloved by dinghy sailors, and Wainfleet in South Lincolnshire, scarcely accessible behind a muddle of sandbanks, which took sea-coal and then provided shipments of corn and malt. Sometimes less usual cargoes came to the Tyne. Chichester sent 50 loads of beech-rails for the colliery tramways which were of increasing necessity to connect the pits to the riverside staithes, and Hull had a regular trade in both wooden rails and sleepers which made up the primitive wagonways. The first recorded mention of these tramways in Britain, and it was an innovation to have far reaching effects, is in a colliery account of 1597-8. They reached the Tyne a little later.

Perhaps the earliest individual of whom we know to deserve the title mining engineer was Huntingdon Beaumont who introduced a number of technical innovations to the North East and built wooden railways at Bedlington, Cowpen and Bebside about 1605.

When in 1603 a complaint from London on the increased price of 'sea-cole' was made, the Hostmen gave as an excuse the longer distance of the newly sunk pits from the staithes as the older outcrops of coal were worked out, although the creation of the cartel by the owners was nearer the truth. The story of the wagonways' development can be traced a step further when a cargo from Hull to Newcastle is recorded to have included "10 coal wagon wheels". As this dates from the early eighteenth century they were no doubt flanged and of iron. Wagons drawn by horses over hard-wood rails or soft-wood ones capped with beech had replaced carts for haulage from the pit-head to the staithes in the mid-seventeenth century. The two fore wheels of the horse-drawn wagon were of iron while the rear pair, against which the primitive brake acted, were of wood.

Despite the competition and restrictions resulting from constricting Elizabethan 'monopolies' there were still some 200 salt-pans in use at Shields at the mouth of the Tyne, although Shields was now referred to as "a station for our coal fleets". Another newly associated industry had been established at Newcastle by Sir Robert Mansell; this was glass manufacture, given a new impetus by immigrants from the Continent. It was one of the first industries to change from using wood

Shipping off Newcastle upon Tyne c. 1690. *Author*

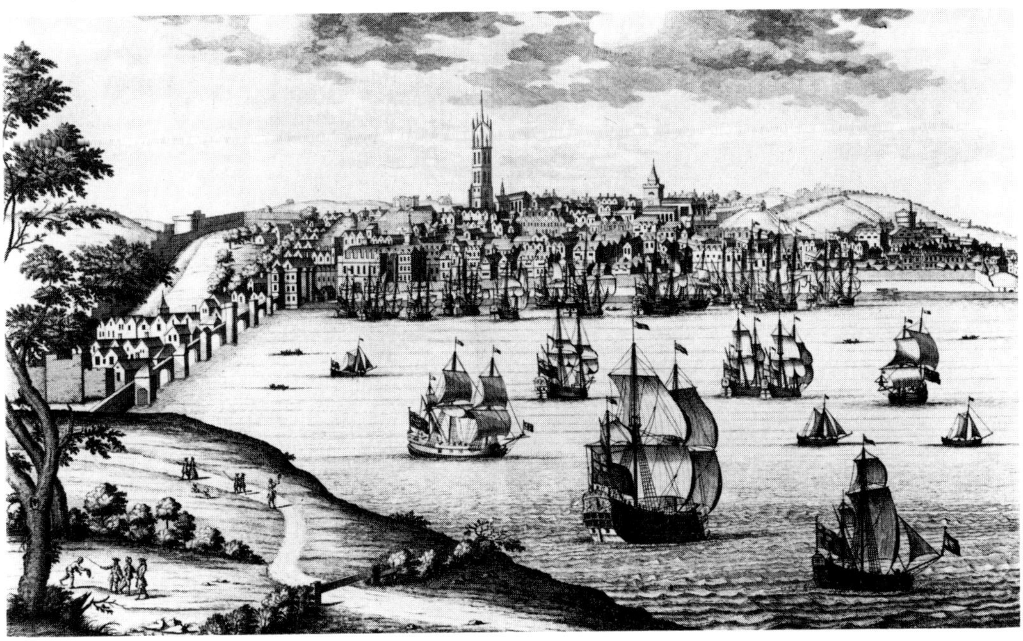

as a fuel — — a change assisted by official prohibition of timber firing of the kilns. After unsuccessful experiments at numerous places, it finally met with success on the Tyne in 1619 and flourished there and at Hartley for over two centuries. It only became important after the Restoration when glass was shipped in bulk not only to London but all along the East Coast. The pack-horses and wagons of the period would have made short work of such a fragile material but transportation by sea and then by inland waterway, made its distribution a practical proposition. It is possible that the ballast cargoes of sand, originally dredged from the Thames and discharged from the returning colliers were conveniently employed in its manufacture. Alum production began, after much difficulty, at this time and used coal heavily in its preparation.

The wars, both civil and with the Dutch, which beset the sea-coal trade in the mid-century, were followed by plague. A fascinating insight into the ambivalent attitude of the age is furnished by two contemporary writers, one looking back to an age of superstition, the other forward to an era of scientific inquiry and technology. Firstly an account of seventeenth century public health precautions. "In the hope of dispelling it (plague-venom) in the atmosphere the Lord Mayor ordered huge bonfires to be lit in the streets, close to the Thames. Coal from New Castle, which was fetching £4 a chaldron, was used to ensure a strong blaze". The price of coal was more usually 14s. a chaldron, an indication of how captains were preferring to seek safe, if less remunerative places to discharge their cargoes. To match this pathetic gesture of desperation we may turn to the inquisitive urbanity of John Evelyn, who, a few years previously, on 11th July 1656, recorded:

> "Came home by Greenwich Ferry where I saw Sir John Winter's new project of charring sea-cole to burn out the sulphure and render it sweete. He did it by burning the coals in such earthen pots as the glass-men mealt theire mettal, so firing them without consuming them, using a barr of yron in each crucible or pot, which barr has a hook at one end that so the coales being mealted in a furnace with other sea-coals under them, may be drawn out of the pots sticking to the yron, whence they are beaten off in greate halfe-exhausted cinders, which being rekindled make a cleare pleasant chamber fire, deprived of their sulphur and arsenic malignity. What success it will have, time will discover."

Indeed, time has; and now, although we have had to wait some three centuries, there are smokeless zones and the fuel produced by Sir John, "deprived of arsenic malignity" is a much sought after commodity.

The ships which brought in the steady half million tons of sea-coal each season to the Thames between the year of the Great Fire and the accession of Queen Anne, were rarely owned by the exploiters of the Northern coal-pits. Yarmouth owned the biggest proportion and Newcastle was, at this time relatively low on the list. Brand in his *History of Newcastle* published a list of ships "useing the Coale Trade at Newcastle" between 1702-4; he numbered 211 from Yarmouth, London owned 168, Whitby 98, Lynn 74, Newcastle 98, Scarborough 54, Bridlington 48, Ramsgate 42 and Ipswich 40. The wealthy coal-owners along the Tyne nevertheless became share-holders in ships, even in vessels owned and registered at other ports. This could arise through chance, when a ship-master found himself unable to meet the full cost of a cargo he had purchased on credit and was forced to part with his share in his command in the absence of any other resources to meet the debt. They might be purchased as a 'hedge', as an investment in an enterprise which was not entirely involved with coal. It was also a method of ensuring that pressure could be brought to bear by a pit-owner upon a captain, when tonnage was scarce, to load at his staithe. Nevertheless there was sufficient divergence of ownership for the clear identification of separate shipping and coal-owning interests in contemporary discussion in and out of Parliament of problems arising in the sea-coal trade.

The major part of the coastwise shipping and the greater proportion of the coal-fleet was invariably owned on the basis of division into divisors of four and the units were frequently widely spread amongst the members of a coastal community. The collier *Lambton Anne* of 200 tons had 17 owners in 1719. No doubt her master would have heartily concurred with the remark of a skipper of one of the last sailing coasters who was similarly placed "that you had to keep a 'going to satisfy that lot" It was, as far as one can judge at this distance of time a system which made it possible for ownership to be spread over a wide range of the social spectrum and afforded an opportunity for a modest investment which scarcely existed until the twentieth century. Generally the owners fell into four categories; there were the masters themselves, these gentlemen seldom held a majority holding and were usually appointed and, if unsatisfactory, discharged, by a group of share owners. Ship chandlers, sail-makers and ropery owners established their right to supply necessities to the vessel by purchasing part-ownership while merchants, as we have seen, could influence the taking up of cargoes, to their own advantage, by becoming investors. These men, by informal contact, in the style of the eighteenth century effectively controlled the policy of the venture, leaving what we should today call management to the captain. Finally shares might be settled upon widows and orphans in the hope, not always realised, that some security of income could be vouchsafed them in an insecure age. Ship-builders were involved, sometimes by retaining a residual share in a vessel that they had built as a speculation and traded until it had found a group who were prepared to purchase it.

All these groups had an advantage over investors in other types of vessel for, with the declaration of war or rumours of war, colliers were much sought after as transports for men, munitions and horses. Fortunes were not made in this way, but it gave a briskness to building and business less known in other trades.

It was common for the master of a collier to buy coal on behalf of his owners and to sell it on their account 'at the Gate', as the coal market at Billingsgate was known. The master was a key figure in the whole undertaking of collier-ownership; treating with the Lords of Coal, arguing with Newcastle fitters, berating keelmen, cursing at lightermen and placating coal-meters with gin and water. While in the last days of the sailing-collier this simple method of ownership was superseded, particularly on the South Coast, by fleets owned by limited liability companies it continued to exist until the end, with masters of tiny ketches purchasing their few score on the Tyne or Humber and discharging it alongside tumbledown quays at half-forgotten ports.

Colliers leaving North Shields after the break up of the 1851 strike. *The Illustrated London News*

5 The Benefit of Boats

"There is not any Town or City, which hath a navigable river at it, that is poore; nor scarce any that are rich which want a river with the benefit of boats."

John Taylor (1580-1653)

TODAY the first impression that a chart of our inland waterways gives to the casual viewer is of a complex and widespreading system. There is a central rectangle of winding and recrossing routes based upon the Midland plain, where it is sodden and unkind, connecting with the Mersey in the north west. Towards the south west is an outlet to the winding Severn and to the north east, by means of the smug and silver Trent, a connection to the fiercely tidal Humber and its numerous tributaries. This complicated pattern is largely a product of the second half of the eighteenth century and the first decades of the nineteenth, the Canal Age, a period in which engineer and speculator flexed their muscles in preparation for tackling the heroic projects of the Victorian epoch. Separate from this system down the length of the East Coast, and extending along the shores of the English Channel, are the trade routes of the navigable rivers, later canalised, and in use since medieval times. These rivers, whose upper reaches were canalised and improved at an earlier time by 'adventurers' and 'undertakers', as the group of local worthies sponsoring such work were known, were used to distribute 'sea-cole' far inland.

Waterways enabled coal to find its way inland on a scale and for a distance scarcely imaginable today. Even now senior inhabitants of Bury St Edmunds, deep in the midst of rural Suffolk, refer to the river Lark as the 'Coal-River', although it is many years since the last cargo, transhipped at Kings Lynn, was unloaded from barges near the town. As in so many cases, legal disputes rather than geographical difficulties bedevilled the completely successful canalisation of the river Lark. The navigable reaches stopped one mile short of the ancient town. The fuel which gave the river Lark its workaday name entered a complicated system of waterways which emanated from the distant Wash ports. Defoe remarked, "that there is the greatest extent of inland navigation here, of any port in England, London excepted . . . . . the merchants of Lynn supply about six counties wholly, and three counties in part, with their goods which has given rise to this observation on the town of Lynn, that they bring in more coals, than any sea-port between London and Newcastle". Indeed, between the Civil War and Defoe's tour in 1724-7, a period of some seventy years, the mileage of navigable rivers had increased from 685 to 1,160, much of the additional distance directly advantageous to Lynn. Long after Defoe's day, in the opening years of the nineteenth century, Lynn was still, by means of the Great Ouse and its collateral rivers, distributing coal to towns far inland at the rate of about 160,000 chaldrons annually.

Very early evidence of Fenland villagers using water-borne coal exists although this has been identified as originating in Nottinghamshire. Coal lent itself to transportation by the then existing net-work of rivers more easily than alternative fuels and this was taken advantage of even in Roman times. In 1585 considerable improvements were made on the Great Ouse, some of the earliest recorded. However, the really effective use of water transportation here did not begin until the seventeenth century. This was when the extensive drainage of this unique area was undertaken by the redoubtable Dutch engineer Sir Cornelius Vermuyden and his Flemish labourers – undertaken primarily to exploit it by agriculture. It had the additional advantage of making the navigable rivers of the Fenland a reliable form of transportation throughout most of the year. Vermuyden's schemes failed to solve all the problems which beset this low-lying area, for its drainage still remained vulnerable to certain combinations of tide, wind, and weather. But when one realises that his efforts were made the subject of political capital in the struggle between King and Parliament and that he worked on in a country torn by sectarian strife, the benefits his labours brought in extending the length of navigable waterways are beyond praise.

As a result of Vermuyden's herculean labour a Fenland lighterman could set out in his beamy double-ended boat, towed by horse and built to twist through the remote Fenland drains, and find his way into the green heart of England. Progressing up the Nene from Wisbech, a self-respecting port with deep-sea ships along its quays, he could reach as far as Wellingborough, passing through Peterborough. Peterborough, I am glad to say, unlike the majority of the towns mentioned in this chapter, still handles water-borne cargoes, but they are now brought by hefty steel motor-barges. It was at this point that Newcastle sea-coal came into competition in later days with fuel brought from the Midland fields by canal. Alternatively, if our lighterman had branched off down the bleak Middle Level navigation, the little town of Ramsey had quays upon which he might land his freight. This was the head of navigation; now the quays and basin are a car-park, a fate which all too often befalls such potential amenities.

Should he begin his voyage along to Ouse the lighter would load coals from a brig at Lynn Docks and, by-passing Denver Sluice, which is the key to the Fenland drainage schemes old and new, he would enter the first tidal lock, still known as the "Dog-in-a-Doublet", after a local lighterman's inn. The fen-lighter, loading up to 40 tons, could then work inland as far as Littleport. This tonnage could encompass sea-going vessels within living memory, as well as canal-lighters. Craft whose holds had been filled beneath grimy Tyneside staithes unloaded their cargoes with less dust, dirt, and altercations, at a countryman's pace alongside little quays bordering the Fenland drains, amidst kingcups and skylarks. Beyond Littleport only smaller, un-rigged lighters could penetrate. Barges loading between fifteen and twenty tons could make their way to St Neots and after about the year 1700 as far as Bedford. This town was only reached by the improvers of the river after much effort and for many years Great Barford, a little lower than Bedford, was the highest point that river-traders might bring their barges. One merchant imported no less than 2,120 tons of coal there in the six months in 1627-8, although much of it must have found its way to Bedford.

The inland voyages made by these lighters were far from uneventful. Iron-prowed barges were employed for ice-breaking in winter, with crews of labourers rocking the boat from side to side to break up the frozen surface of the river. In the centuries before adequate pound-locks, guile and muscle-power had to be employed in order to create sufficient depth of water to float loaded craft through the shallow reaches. The lighters carried aboard 'staunching tackle', designed to assist in producing a temporarily deeper stretch of water. This consisted of wattle hurdles, substantial posts, and canvas sheeting with which a temporary dam could be built. As a short-term expedient an unladen lighter would be moored athwart the stream to create an artificially high level of water. It speaks much for the construction of these craft that they could withstand such robust treatment and survive.

By the nineteenth century the lighters moved in trains of five or six, towed by a steam tug and joined together, not with conventional tow ropes but by wooden 'sprits' held by 'fest ropes' and with the stems of the lighters close coupled by seizing chains. The lighter immediately astern of the tug was steered by an extra long 'sprit' or steering pole. The rigidity of the substantial wooden poles between the others of the train prevented over-riding when the little convoy either slowed down or stopped. When the procession turned a sharp corner someone had to run from lighter to lighter, letting go the 'fest' ropes so that the lighters could swing; no rudder was used. The helmsman steered by pushing on the bowsprit of the second lighter while he stood on the stern of the first. The whole process must have kept the blood circulating even on the coldest of Fenland winter days.

One lighter in the train was always a house-lighter, built with a stern cabin, which was fitted with a stove, the nature of the cargoes ensured no shortage of fuel, and so provided some comfort for these Fenland water-gypsies as they travelled the bleak levels and lodes. In the days of horse towage the horses were trained to jump the fences which met the line of the waterways and were built to prevent stock straying; these were deliberately restricted by law to 2 ft. 6 in. in height. The horses were also taught to ride aboard the lighters when the wind was fair and a primitive square sail could be set with advantage to blow them along a level.

The furthest point to which water-borne trade could penetrate into East Anglia from an inlet at the Wash was Bury St Edmunds which was receiving sea-coal as early as 1687. Thetford could be reached by loaded barges working on the Brandon River, made navigable in 1670, during the widespread upsurge of speculative enterprise after the conclusion of the Civil War. If one visits Mildenhall, another minor Suffolk country market-town, evidence of a dependence on water for conveying its bulk imports is still to be seen in the siting of the town's miniature gasworks. It stands, now a little forlornly, on the waterside, not, as one might expect, near the now abandoned railway lines. Further south, Cambridge evokes many images to the reader but that of an inland river-port is not usually amongst them. Yet the Common Hithe was once on the riverside opposite Magdalene College and King's Mill was the head of navigation. Before the sluice at Denver was built in 1650 the Ouse was tidal for considerable distances inland and small sailing craft, built and rigged for trading about the Wash and even beyond, reached Cambridge. In later days it could take as long as five days to reach the town from Lynn, but until the very end of the Cam's commercial employment, terminating with the arrival of the railway at Cambridge in 1845, Newcastle coal was its principal import while wheat filled the barges for the return journey.

An unusual feature of the last stage of the voyages to this inland town was that bargemen's horses plunged into the river and towed along a sunken causeway, with only their heads and withers showing above water. Today sloping lawns reach down to the river and of the hythes, warehouses, and busy quays nothing remains. Now only summer punters sometimes feel the causeways with their poles where once the towing-horses found a precarious foothold.

Map to show the rivers and waterways employed to distribute seaborne coal in the Eastern Counties.
Map by Roger Finch

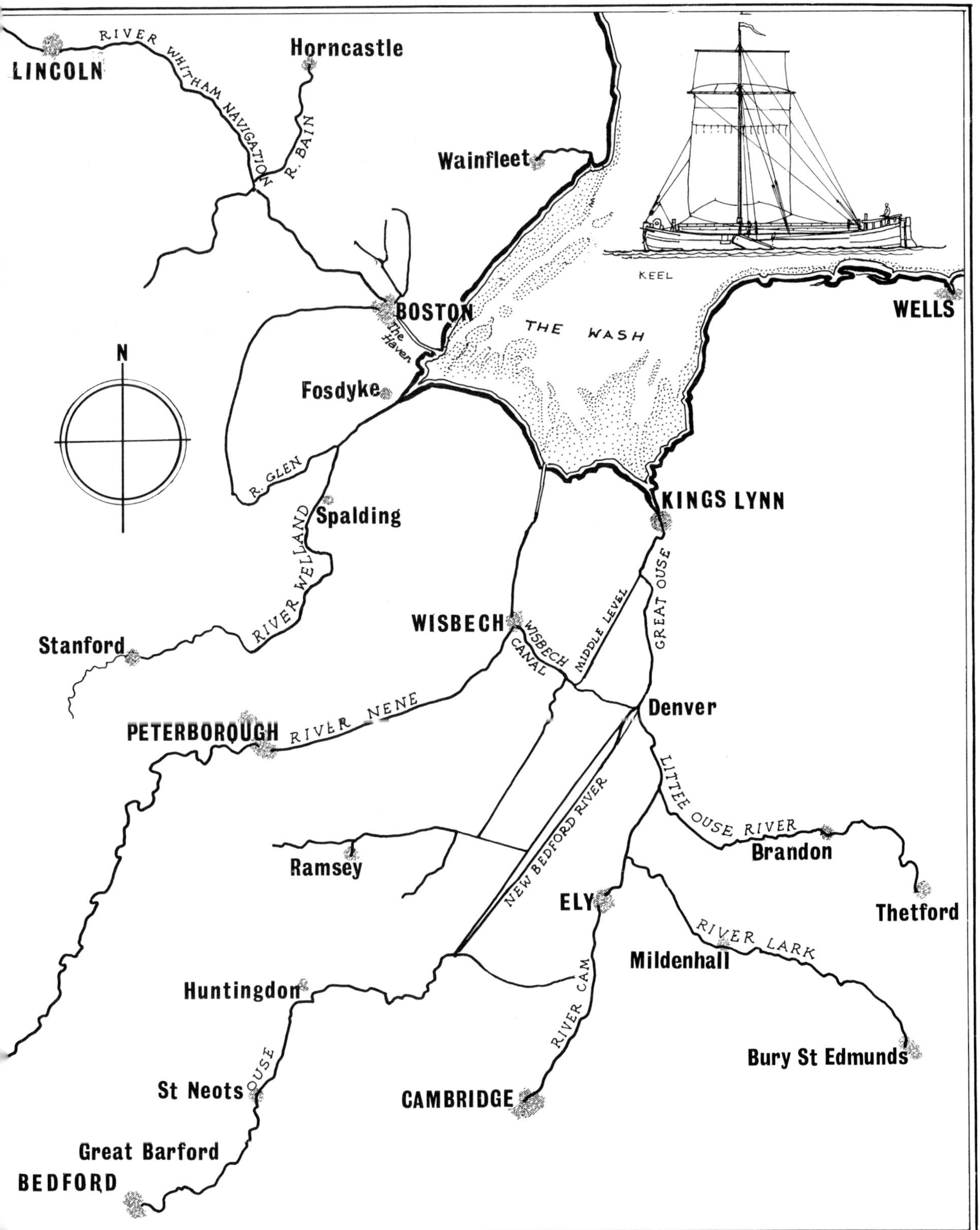
LINCOLN
RIVER WHITHAM NAVIGATION
Horncastle
R. BAIN
Wainfleet
KEEL
BOSTON
The Haven
THE WASH
WELLS
N
Fosdyke
R. GLEN
Spalding
RIVER WELLAND
KINGS LYNN
GREAT OUSE
WISBECH
WISBECH CANAL
MIDDLE LEVEL
Stanford
PETERBOROUGH
RIVER NENE
Denver
LITTLE OUSE RIVER
NEW BEDFORD RIVER
Brandon
Ramsey
Thetford
ELY
RIVER LARK
Mildenhall
Huntingdon
RIVER CAM
OUSE
Bury St Edmunds
St Neots
CAMBRIDGE
Great Barford
BEDFORD

Retracing our steps to the borders of the Wash and crossing over to the north-western corner we may enter a waterway system which once brought lighters to Market Deeping, and a considerably older one beginning at Boston. This later system joined the Witham river to the Trent and passed through the Fossdyke to Lincoln. Perhaps the most ancient artificial cut in the country, it may be of Roman origin; we are certain that as early as 1121 Henry I authorised its rehabilitation, and the wool trade of the Wolds must have found it particularly advantageous. Later various bodies bewailed the inadequate state of the river Witham; in 1671 it was said to be "much silted and landed up and thereby not passable with boats and lyters as formerly, to the great decay of trade . . . . ". But eventually dredging was done and the banks cleared, for a century later we read that the Witham was once more "a good navigacion betwixt the burrough of Boston and the river of Trent". Fortunately for posterity this was just in time for the English water-colourists of the early nineteenth century to find a favourite subject in a laden coal-barge passing under the pleasing medieval decay of the High Bridge at Lincoln.

The form of the Fen lighters was primitive; they were essentially big open boats, slab-sided, hard chined with no pretensions to grace, evolved to carry the maximum tonnage on the shallowest draught, unadorned and unloved. The keels of the river Witham were an altogether different class of vessel; to see one in stately progress on the river to Boston, or lying in the Lincoln Braybrook with the cathedral in silhouette beyond the mellow brick warehouses, was to be transported back to the Middle Ages. They were capable of crossing the Wash and the holds were sealed with hatch-covers; they had massive tillers and a single mast from which was set a large square-sail of surprising efficiency. On the fore-deck was a windlass of the same primitive type that the sea-going colliers carried and the whole appearance of the massive bluff-bowed hull was that it derived from the same ancient mould as the nineteenth century barks and brigs built at Newcastle and Scarborough.

I well remember seeing a group of keels at Thorn, on the Stainforth and Keadby canal, at the old Sheffield and South Yorkshire Navigation yard. They might have been a fleet of medieval craft, newly returned from the battle of Sluys. For one familiar with West Country schooners, ketches, and Thames spritsail barges it seemed that nautical evolution had passed them by. Their construction was massively archaic; the timber heads for belaying warps, the wooden rails at the stern, the almost semi-circular bows, and the cavernous holds seemed to belong to another epoch. The descent from the hatch to the fo'c'sle was effected, not by a ladder but by a notched vertical pole! For all their grace the finest Thames sprittie cannot disguise the fact that she has grown up from a slab-sided river-lighter. Even without masts and rigging, far from the sea, benighted in an industrial no-man's land, these keels had an indefinable air of having descended from medieval royal nefs and balingers. One can only regret that the efforts to preserve and maintain a keel sailing the northern waterways, have, so far, met with only a limited success.

The square-sailed keel and its sloop-rigged variant was essentially a vessel of the canals and waterways of Yorkshire, all connecting with the central artery of the tidal Humber. They worked between the Humber ports and Selby, where many were built, Goole, the Market Weighton Navigation and up the Don and the Aire and Calder. This extensive network was, however, only involved in the distribution of Tyneside coal on a modest scale and then only until Stuart times. The developing Yorkshire coal-pits rapidly took advantage of waterways originally improved and canalised for the benefit of the North Country woollen industry. The first of these navigations to be promoted is still one of the country's most effective commercial river improvements, in use today and still profitable; this is the Aire and Calder. The imposition of taxes on sea-borne coal in 1695 made this long-discussed project of improvement doubly attractive as land-sale coal carried no such imposition. The collieries nearby, upon the completion of the work could seek an outlet for their production at York and other towns who had previously relied upon Newcastle coals laboriously and expensively transhipped at Hull.

The succeeding years saw the navigation promoters hard at work elaborating and deepening an already widespread river-system. This was supplemented by tramroads and waggonways which brought coal to the river quays from the pit-head, somewhat in the manner of the rail-links serving the Tyne staithes. But there was little, if any, attempt to thrust seaward into direct competition with the Newcastle factors and the long-established sea-borne coals from the more northerly ports to London. Export to the Continent was even rarer. Although Goole was bravely developed in the eighteen twenties as an inland port for sea-going vessels, the tonnage of coal shipped coastwise was negligible. Unfortunately for its promoters it was opened only a year before the removal of taxes upon the Newcastle coastwise trade in 1831. Immediately before the railway mania of the 'forties Goole's coastwise trade in coal amounted to less than point one per cent of the North East's two and a quarter million tons, while even less went abroad.

The little coasting sloops and ketches of the Humber, known as billy boys and and really a sea-going adaptation of the keel, rarely exceeded eighty or one hundred tons burthen. They traded regularly to Louth, Boston and Kings Lynn with coals, but their owners could not break into the tightly controlled London market with its tight ring of factors and dealers. If they came into the Thames estuary it was more frequently with building stone than coal. Only when railways and mechanised loading, cheap steam tonnage and reliable steam towage were available did the Yorkshire and East Midland pits find an adequate outlet, particularly with gas-coals, to the sea and the coastwise trade to London. Eventually, however, it was from Keadby and Goole, that in the nineteen fifties the very last cargoes of sea-coals were loaded to be sailed along the colliers' highway to the Thames estuary ports by the last of the sailing coasters, the big mule-rigged spritsail barges.

The keels and billyboys, during the eighteenth and nineteenth centuries, worked about the Wash and its rivers which formed one system for distributing sea-coal far inland. Further south, along the coast of East Anglia were other waterways which, by the time that the railways had largely superseded them as highways of trade, had been effectively improved for inland navigation. So extensively was canalisation and improvement promoted that at the height of their usefulness no village in the Eastern Counties was more than fifteen miles from a river-quay or wharf.

The grandiloquent prose of the preambles to Stuart Parliamentary Bills and renewal of patents seeking the right to improve waterways have led some authorities to imagine that they were, in every case, followed by immediate and energetic action. Unfortunately this was not always the case and rarely does work on the projects date from this time. The East Anglian Stour was mentioned as early as 1638 as being canalised, but in actual fact it was 1708 before work commenced "to make the River Stour navigable from the town of Manningtree, in the County of Essex in to the Town of Sudbury in the County of Suffolk . . . . . effecting a passage of barges . . . . . beneficial to trade and advantageous to the poor". For advantageous to the poor these early navigations certainly were. When the Chelmer, also in Essex but further south, was canalised after many false starts in 1797 and the first brig sailed into the basin at Heybridge, it had an immediate effect upon prices. To quote from a contemporary advocate of the project . . . "The price of land carriage for coals brought by wagons from Maldon to Chelmsford is 8s. per ton, whereas the tonnage and lighterage by water may be charged at 2s. a ton, to which adding a toll of 2s.6d. the price will be 4s.6d.".

A few years before, the river Gipping had been made accessible for horse-drawn barges as far as Stowmarket, linking it with the estuary at Ipswich. This brought the possibility of a cheaper carriage of bulk goods to the rural area north of the Stour while further up the coast the Blyth served another part. The river Blyth, whose estuary entered the North Sea between the lost medieval port of Dunwich and the town of Southwold, was turned to commercial account in 1757. Work then began on the locks which enabled the first cargo-carrying craft to reach as far as Halesworth four years later, when the work was completed. A local journalist recorded the event; "This day we had the pleasure of receiving into our bason a keel from Southwold laden with coals drawing upwards of three feet of water . . . . . the barge was attended from the Town Lock up to the great bason by a numerous concourse of people . . . . . ". Sixty years later the port of Southwold

A model of a trading hoy c. 1750. These shallow-draught traders brought coal to inland ports and small coastal havens. *Hugh Moffat*

was importing some 10,000 tons of coal annually in its little fleet of sailing colliers, much of it proceeding by barge up the Blyth and thence by road to the surrounding countryside. Malt, corn, and barley were exported to the northern coal-ports in the brigs and schooners which brought in-coming cargoes. It was a hazardous trade and the register of many of them terminates with a curt "wrecked at Winterton" or "Lost on Hasbro' Sand", cloaking a grim drama with few witnesses.

Yarmouth competed with Kings Lynn amongst the out-ports as the greatest importer of sea-coal. At the height of the coastal trade under sail, of the 45 ports outside London which landed Newcastle coal these two and their associated 'creeks' alone accounted for half of the trade. At Yarmouth in 1714, 30,000 London chaldrons were imported into the haven, but only one-sixth were landed at the town's quays. Just over a century later, immediately before the advent of serious railway competition, the figure had trebled.

The smaller coasters, those loading less than about 100 tons, could and did proceed up the winding Yare on to the Wensum and thence to Norwich. However, the majority, the brigs, brigantines, and barks went no further than Yarmouth. There they lay along the Yare in picturesque lines moored so that their bowsprits overhung the quay to which Defoe referred to in such glowing terms ". . . the finest quay in Europe, not inferior even to that of Marseilles itself". The coals that were not destined for the town itself were discharged for transhipment into keels, as the sailing lighters of the area were called. They bore only a superficial resemblance to their grimy namesakes of the northern rivers. They were, on average, much larger,some being capable of loading as much as 80 tons and had living accommodation built in the bows for their two or three man crews. These keels of the Broadland rivers were more obviously descended from the clinker built Saxon *knorr* of the ninth century and were built with a pleasing sheer and flaring bow; like their equivalent of the Tyne, the hold was open and they carried a simple polemast amidships from which was set a square sail. Scores of these craft sailed, towed, quanted, and tracked their weary way to Norwich or sought out the now neglected red-brick coal stores which still stand on the village staithes. They went up the Waveney, as well, as far as Beccles, returning with barley and malt.

When conditions of tide and weather permitted it, a loaded collier would anchor off in the Roads. Then keels would sail out and the cargo would be discharged into their open holds while the anxious captain of the brig watched the sky for signs of deteriorating weather and the harassed mate shuddered as the rope fenders between his ship and the keel alongside ground to shreds in the sea-swell. It is difficult, at first, to understand how such a potentially dangerous situation was tolerated. The reason lies in the fact that, like the Northern coal ports, the East Anglian harbours had entrances which shoaled and altered alarmingly, particularly under the stress of a series of winter northerly gales. The eighteenth century, with its own agricultural revolution gaining momentum, saw a further aggravation of the problem unintentionally brought about by the intervention of enclosing land-owners. Their reclamation of the tidal marshes along the banks of river estuaries significantly reduced the scouring action of the ebb-tides by cutting down the amount of water absorbed on the flood. This scouring action contributed much towards maintaining deep water at the harbour mouth which enabled vessels to enter and leave with a reasonable margin of safety; once it was reduced below a critical level, results could be catastrophic to the usefulness of the haven. Before modern methods of harbour stabilisation and maintenance were developed and capital was available to employ them, alarming changes in the depth of water and even position could occur. This left no alternative to the captain of a deep-loaded vessel but to await spring-tides or lightering at least some of his cargo outside the safety of the harbour into barges. Of course, such lightering meant that harbour dues and taxes might be avoided, but this was countered by the keels demanding a higher freight for such hazardous work on the open sea.

The Yarmouth keels did not return empty to the Roads after they had discharged their coal cargoes on the quays or Broadland staithes. They loaded sand at a conveniently placed shoal near the mouth of the river Yare which the tidal flow unfailingly "year in and year out" replenished despite the many thousands of tons removed annually. Few of the colliers could safely make sail without first loading ballast and a steady supply was essential to ships lying empty beside the long quay or discharging in the exposed roadstead.

By the second half of the nineteenth century, before the coming of the railways, if not earlier, the keels had been replaced by the more sophisticated wherries. It is probable that the improvements to the harbour made lightering less necessary and the big, robust keels evolved for this became redundant. The wherries were rigged with a gaff-sail and were, generally speaking, smaller, but more efficient under sail, than the keels. Perhaps their success was due to their having been developed for passenger-carrying and the requirements of keeping to a published timetable, at least nominally. While they never attained the size of the keels, their greater speed and their ability to make a fair wind of it when the keels would have to moor or tow, gave them the edge over the older craft. They survived until the twentieth century and until after the First World War a gas-works at Norwich continued to get coal by wherry, loaded from steamers at Lowestoft.

Lowestoft, a few miles to the south of the much older port of Yarmouth, was created as a commercial harbour in competition with its northerly neighbour at the beginning of the nineteenth century. It failed lamentably at first. The southerly drift of shingle across the mouth closed the entrance to trading vessels of average tonnage almost as soon as it was constructed. This entrance was a man-made cut, penetrating the bank which separated the North Sea from a series of waterways behind Lowestoft, known as Lake Lothing. These in turn were engineered to provide a back-door, by way of the rivers Waveney, Yare and Wensum to Norwich and its extensive trade. Despite the dredging of the entrance it rapidly became so shallow that it was impossible for a loaded collier to use the quays and warehouses optimistically planned along Lake Lothing. However, the story had a happy conclusion for, despite its failure to become a port for Norwich, Lowestoft ultimately developed as a fishing station.

Before turning to the various schemes, successful and otherwise, which were made in an attempt to use the modest rivers of the South Coast as highways for trade, we might conveniently link it with the Thames where the colliers brought so much of the sea-coal. Some had always been transhipped from colliers as they lay at Limehouse and taken up the Lea to Hertford, for an Act to 'scour and amend the River Lea' was passed in 1430. When the country towns of Ware and Stortford were made accessible to water-borne traffic in the late eighteenth century 50 barges were employed in the coal-trade, sailing and towing between the market-gardens and cabbage fields to Broxbourne and beyond.

It was not until the early part of the seventeenth century that coal was able to ascend the Thames to its tidal limits. Successive obstructions delayed improvement; wily millers, fish traps on stout wooden frames spanning the river, shoals and dams, all were hedged round with legal defences in depth and ancient rights. No worse perhaps than on many other waterways, but effective in standing in the way of the river's full use as a water-highway. But progress was made and by 1605 it was possible to reach as far as Burcot, near Oxford, although the journey was by no means easy. Old London Bridge, at this time, presented a barrier such that "it is so much a mystery and profession to navigate through the bridge that the craft of the inland navigation cannot venture into this complicated danger". One of the numerous lost opportunities proposed after the Great Fire of 1666 was to link Bow Creek, below Greenwich, with the river Fleet, so that traffic might avoid the cataracts rushing between the piers of the bridge while London would be circled with a commercial waterway and defensive moat.

The commercialisation of the waterways which joined the Thames had to await the really successful improvement of the major river. Most important of these was the Surrey Wey which was surveyed and for which a scheme was produced for making it navigable in 1612; but it was not until the 1650's that work on the river made it possible for barges to reach Guildford.

The Wey reaches the Thames immediately below Shepperton Lock and despite disruptions in the work for its improvement inevitably caused by the Civil War, it proved highly successful. A diversity of rural produce, ranging from cabbages for Covent Garden to timber for the royal shipyards at Woolwich, made its way along the river, while sea-coal, loaded in the Pool of London provided a staple return cargo. Eventually, during the canal boom of the late eighteenth century, the navigation was extended beyond Guildford to Godalming, further, the Basingstoke canal was constructed, branching out at right angles to the line of the Wey, towards the west. The charm of the scenery through which these waterways pass is, alas, an indication of their extreme vulnerability. There were no industrial developments to furnish cargoes except for the last remnants of the Whealden iron smelters. At the first chill of approaching railway competition and the resultant availability of coal from the Midlands, their profitability, never high, sank out of sight. The lower reaches of the river Wey were more fortunate and were used

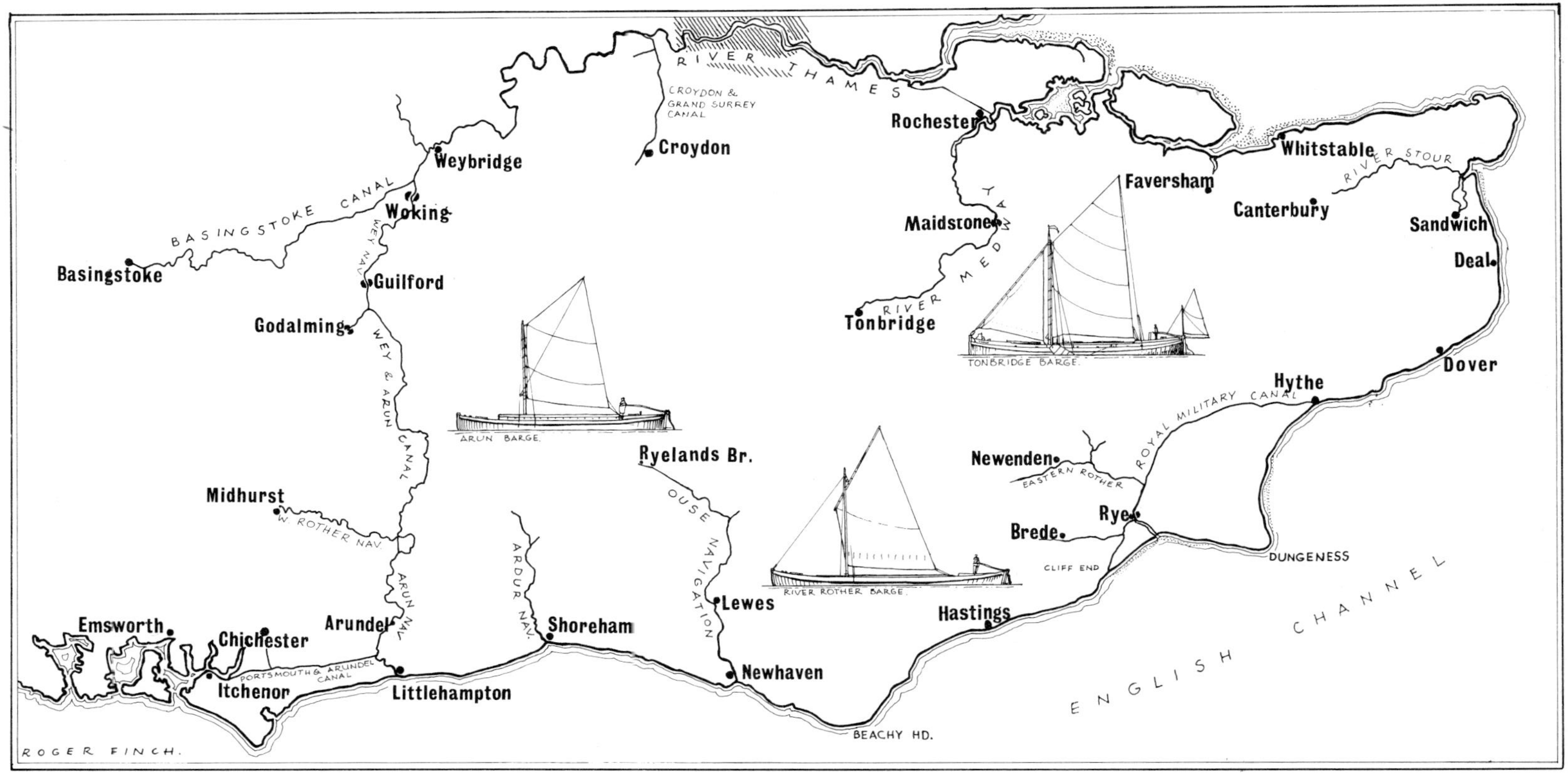

Map to show the rivers, waterways and canals employed to distribute seaborne coal in the South East.
Map by Roger Finch

commercially until 1958 and as far as Guildford until 1969, coal featuring amongst the cargoes until very nearly the end. The barge owners of the 'agricultural canals' did their best to compete with the alternative presented by the railways from the 1850's onward. This is clearly demonstrated by the steep downward trend of coal rates on the Wey and Arun canal, coal being the largest proportion of the cargoes carried. The rate was 3d. a ton-mile in 1822, dropping to 1½d. in 1844, when the iron roads were building, then to 1¼d. in 1849, and finally, as the commercial end of this particular waterway was in sight, to 1d. in 1850.

The Wey and Arun canal was, as the name indicates, an ill-fated attempt to provide a connecting link between the Thames, via its tributary the Wey, with the river Arun which flows out into the English Channel at Littlehampton in Sussex. The Arun is one of the many streams of the South Coast which, while modest in scale and only moderate in navigable depth and length, helped to bring coals inland from the South Coast ports. The wide Medway is an exception to these, both in size and use and to which we will return later; here we can re-create a picture of little lug-rigged lighters working as far inland as the village of Newenden in Sussex with coal loaded from Rye's big coasting-barge colliers. On the Sussex Ouse, where the barges were spritsail-rigged, the town of Lewes had long been accessible by water; indeed large coasting vessels had been built there during the nineteenth century from Sussex oak. In the 1800's the navigation was extended some twenty miles, but inevitably the building of the coastal railway line brought about its end, although to this day small motor coasters still work up the tidal section to Lewes.

Nearby Shoreham also had its collier fleet. A proportion of their cargoes was transhipped up the Adur which had been cleared with the original intention of land drainage, and at the very end of the canal age it was lengthened by a few miles as far as Baybridge. Somewhat earlier, as part of the ambitious scheme to form an alternative to the sea-route between London and Portsmouth a grandiose plan had been initiated by a group led by the Earl of Egremont. His Lordship stands out amongst the over-sanguine promoters, hopelessly optimistic surveyors, shady financiers, and inefficient contractors of the contemporary schemes as a man of integrity and principle. His life-style may have been one that we associate with the Regency period in which he lived, but this concern with the deserving poor and boundless energy in the promotion of large-scale enterprise was unmistakably Victorian. He was involved with the Wey and Arun canal and then, together with others, in the promotion of a proposed link between Portsmouth and a route which would take traffic to the Thames and thence to the Pool of London. Sea-coal had for generations been discharged from coasters into lighters at Itchenor and taken five miles to Dell Quay where it completed the journey to Chichester in the traditional broad-wheeled horse waggons. The canal finally to link London to the South Coast, so dear to Lord Egremont, began at Ford, on the Arun below Arundel, then joined a bigger and much deeper section, capable of taking coasting vessels, that ran from Chichester Harbour to Chichester and to the estuary at Birdham. The barge traffic was then towed or sailed along a dredged bargeway amongst the mud-banks of Langston Harbour and through a cut across Thorney Island; this channel terminated in a basin at Salterns Lock. The route was inaugurated in 1822 by the schooner *Richmond* discharging a cargo of coal at the basin end of the canal, amidst the traditional plaudits of the attendant wellwishers.

But the idea of by-passing the North and South Foreland by canal and 20-ton barges, which might possibly have had some validity during the Napoleonic Wars, had little chance of implementation once peace came. Today the lower reach of the Chichester Canal, which had its origin in this pipe dream, is used by yachts for laying-up and repair, while cows graze in the shadow of the downs along the dry bottom of the section intended to take barges to the Arun.

There were schemes in Hampshire as well as Sussex and by making several artificial cuts, the river Itchen was made navigable to Winchester by the mid-eighteenth century. Canal barges loaded cargoes at Northam which was as far as the coasting sloops and brigs could reach. Above nearby Southampton, a route led to Andover, following the course of the river Test, while a branch led off towards Salisbury, winding through the chalkland hills. All faded and died as the threat of the railways became a reality in the 1850's.

Writing generally of the cargoes that were carried and the variety of goods loaded from the rural quays which served the waterways two centuries ago, it can be said with a fair degree of certainty that coal never ceased to be less than half of the total carried for any one year. 'Deals', as soft-woods were known, dominated the rest, while a rich diversity made up the total; glass bottles and eggs, (short work would have been made of them by the jolting waggons on nineteenth century roads), tar and tallow, bark for tanning, and seaweed for fertiliser. The mill-owners, stubbornly refusing co-operation, were sometimes placated by canal companies carrying their mill-stones aboard barges free of charge or at a reduced rate. As time went by and the promises to shareholders of more valuable cargoes seemed less likely to materialise, coal came more and more to dominate the receipts of the barge-owning companies.

While the railways rapidly eroded the limited profits of the minor waterways constructed with such high hopes, the trade of the Medway was in a different category. Originally this Kentish river's improvement as far as Tonbridge had been stimulated by the necessity of discovering an economic means of bringing timber to the Navy building yards at Chatham and Sheerness on its estuary. But this trade in strategic materials outwards was, from the mid-eighteenth century, principally balanced by coal shipped inwards as far as Tonbridge. Unlike the other canal-owning companies, the proprietors of the Medway navigation extended their enterprise to trading with their own craft, although this was expressly against the law as it was considered to create a monopoly. Undeterred by the likelihood of legal action they integrated their enterprise further by buying coal by the shipload. There was a succession of improvements to the waterway and the only problem which the proprietors found intractable was the impressment of bargemen by the Navy. The proximity of the naval base made this a growing hazard and despite strong representations to the Board of Admiralty they found protection as difficult to obtain as the keelmen on the Tyne.

No doubt the rising prices of coal and the growing uncertainty of its delivery as colliers were commandeered and crews impressed during the French Wars, encouraged the canal proprietors to buy coal direct from Newcastle factors, freighting it to the Medway and finally in 1804 buying their own collier. This and similar enterprises enabled the company to pay a dividend of more than 20 per cent in successive years, unheard of by other South of England canal investors. But even their lively approach could not compete successfully with the all-conquering railways and by the mid-century the game was almost over, although the company was not wound up until 1911.

Maidstone could be reached through Allington Lock just above the tidal part of the river and spritsail barges, the vessels traditionally associated with the Medway, could lower their masts, pass under the bridge at Rochester, and proceed far up the waterway. During the whole of the nineteenth century coasting spritsail barges from the Thames and the Medway voyaged to the Tyne and brought coal to the Medway towns and villages.

6 The Best Sun We Have

"The best sun we have is made of Newcastle coal"
Horace Walpole, 15th June 1768

COAL extraction in Britain reached the level of one million tons long before any other product attained such an astronomical level. By the beginning of the eighteenth century the volume of solid fuel carried in the holds of the wooden colliers past Gravesend on the Thames, or to the outports, attained an annual level of half a million tons; London imported more coal than any port in the world. There was a falling off from a steady annual increase by the turn of the century when consumption by the 'Great Wen' was affected as the rise in population stabilised. Shipmasters, who at this time bought their cargoes and sold them for as much as they would fetch, were clearing, it was claimed, only ten pounds a voyage, during this depression "even supposing he come and goe well".

But this was only a temporary halt, ending about 1710 and was balanced to some extent by the fact that Newcastle coals went increasingly coastwise as far west as Exeter and even Guernsey. At Exeter they came in direct competition with the South Wales fields and the production of the ancient coal pits in the Somersetshire Forest of Dean. But the shippers of coal from these areas were faced with the expense of cross-country transportation and then the difficulty of their vessels 'rounding the land' if they were to penetrate more than local markets. Colliers, setting out with a fair wind down the Bristol Channel and then weathering the danger of the long Atlantic-roller washed Lizard, found themselves faced with a hard thrash to windward along an iron-bound coast before they could reach a port in the English Channel. Besides this, the system of grading fuels was adopted in 1766 so that those Tyneside pits raising coals of inferior quality could compete with newer, higher quality mines elsewhere. The renown of the

The Nene brought sea-going vessels to Wisbech and from them the Fen lighters loaded their cargoes. A group are clustered around the snow *Vigilant* built at Stockton-on-Tees in 1848 by Craggs. The photograph, dating from 1856, also shows the Portuguese brig *Amelia* in foreground. *Wisbech Museum* favoured house-coals of Wallsend also gave an edge to the Northern shippers. The price the collier masters could demand more than compensated them for the long haul down the length of the North Sea and the English Channel. There was moreover the added incentive of an important non-domestic market in Cornwall, which grew throughout the eighteenth century with the increased employment of the newly devised Newcomen steam-pumping engine installed in the tin and copper mines. Coal used in these primitive devices was free of tax. Coal landed on London quays carried a tax to rebuild the City and St Paul's after the Great Fire and later to pay for the French Wars.

But it was the Thames-side brewers and lime-burners, 'soap-boylers', the potters of Fulham and the brickmakers of Grays and above all the Londoners themselves who brought colliers from the North East by the score to anchor in the Pool and Limehouse Reach. The wealthy purchased it by the ton and stored it in the cellars of their newly-built Queen Anne houses against winter when the Thames, frozen for a mile below London Bridge, prevented even lighters moving. The poor bought it by the stone from hawkers who cried it in the alleys and courtyards of Southwark and Moorgate.

By now the sailing collier had evolved into a distinctive and characterful type, principally built by the rapidly emerging ship-builders of the North East Coast for the London and Baltic trade. These vessels were, on average, much longer and heavier than the general run of contemporary vessel working coastwise in the coal trade. The smaller coasters were usually less than 80 tons and they were small by necessity. Loading for tiny ports at the head of navigation reached along winding, undredged rivers where a depth of much more than a fathom was only experienced at spring tides, a larger vessel would never have attained its objective. It was also common for merchants to expect the colliers to discharge cargoes upon open beaches when to delay for longer than a few tides courted disaster. They went, for instance, to Heacham, 14 miles north-east of Lynn, "where there is no harbour for shipping and small craft unlade their cargoes on the sands", or Cromer, "where vessels of 60–100 tons burthen lie upon the Beach". At Melton, far up the Suffolk Deben and where the channel could be too shallow for an open-boat, a small collier came, and here one of the crew lost a token from Berwick-on-Main Colliery, a souvenir recovered a century and a half later from the river-mud. Although an ancient port and for long an importer of coal, Blakeney in Norfolk, had to wait until it was 'improved' by virtue of an 1817 Act of Parliament. After the improvement it was proudly announced "that vessels of 150 tons can now reach the quay where the spring tides rise 10 ft." although even this was barely enough for the general run of smaller colliers; here the little sloops were at home with their cargoes of 20 or so chaldron of Hutton or Benwell coals. Those who delivered their cargoes on open beaches or in primitive drying harbours made sure of their line of retreat by laying out a kedge anchor. This was done by prevailing upon the wagon-owner who had come down for the cargo, to load the anchor into his cart and deposit it at a suitable distance from the vessel.

The trade in coals to the open beaches, usually for domestic use, survived until the end of the nineteenth century and even beyond, long enough for the camera to record it. Of equal importance in the eighteenth century was the shipping of fuel destined for the lost landing places along the Yorkshire coast which served the nascent alum industry. Alum, much in demand for dyeing and tanning, was produced in north Yorkshire from the beginning of the seventeenth until the mid-nineteenth century, leaving scars still visible on the majestic cliffs. The essentials for this primitive chemical industry were the alum shale, found in the sea cliffs, kelp and urine. Kelp was prepared by burning sea-weed and its collection was a minor coastal industry, like the combing for sea-coal, along the Yorkshire and Durham beaches. Urine was a source of ammonia and was collected from local towns, but there is evidence that this, like the kelp, was shipped to the manufactories.

A water colour of Stone Bay, Broadstairs, painted in 1838 by J. D. Harding. The brig is lying on the shore ready to discharge her cargo into carts. *W. Lapthorne*

The whole industrial process was, however dependent upon a readily available and inexpensive supply of coal while the remote nature of the activity, strung out along a forbidding coastline, made the sea the only possible link with the Tyne or Sunderland. These limitations meant that small sloops of shallow draught and of especially robust construction, whose wreck or damage would not represent an overwhelming loss, were employed to bring in fuel.

The first quarries to be exploited were at Guisborough and the sloops with their few keel-loads of coal, discharged their cargo into carts on the Coatham sands or the river bank, inside the Tees estuary. These rudimentary landing places were havens of peace compared to the inhospitable cliff-foot which represented the only discharging place available for the later development of the alum industry, sited as they were between Saltburn in the north and Ravenscar in the south, along some of the most spectacular of all our coastal scenery. Few more dangerous places could be found for regular use by the little forty-foot merchantmen of the period. Sandsend, north of Whitby, provided some natural facilities and a pier was built out, almost alongside the buildings housing the leaden pans in which the alum was produced. There was another sandy landing at Kettleness, where the harder sandstone produces headlands which give some shelter. While an inlet at Skinningrove was convenient for the Lofthouse works it meant that transhipment into carts was unavoidable. This extra cost was saved by cutting through the rock to form a deep enough channel to float the sloops to cliffs directly below the works. The expertise of the alum miners was also used to blast a dock at Ravenscar in the rock while even the sinking of mooring posts and the construction of staithes required an enterprise and energy which anticipated the achievement of a later age.

The little ships which brought the coal not unnaturally spent as short a time as possible delivering their cargoes on the shore; they stood off for the hours of darkness and returned next day only if the weather remained propitious. At some landings the precipitous cliff-path defied even pack-horses to deliver the cargo and then the expediency of driving a long inclined tunnel from the quayside to the manufactory was used. The coal was hauled up it by a horse-operated gin-wheel and the alum, which sometimes furnished a return cargo, lowered down. However, the value of the alum encouraged manufacturers to accumulate the the chemical in warehouse and then ship it from safer local harbours in larger, more seaworthy vessels.

The 'alum-sloops' were not confined to this dangerous activity and occasionally participated in the corn and general coast-wise trade. One, the sloop *Little Henry* was beached at Kettleness and became the participant in as dramatic event as ever occurred along the whole coast. Shortly before Christmas 1829, a dull roar above heralded a landslide which started in the cliff over the works and soon the entire settlement, storehouses, cottages and manufactory was overwhelmed. The workers

The shape of the eighteenth century collier's hull. The *Edward*, built in 1797.
National Maritime Museum, London

and their families, with such pathetic belongings as they could hastily snatch from the ruins of their homes, took refuge in the only shelter available, the *Little Henry,* now afloat at the cliff foot. With the making of the tide she set sail and with her distraught human cargo safely made port.

The little fellows who sought out the smaller ports were usually, in the seventeenth and eighteenth centuries, single or two-masted, rigged and sparred in a variety of ways with certain regional differences. They were predominately rigged with tall, narrow squaresails, as these set well, particularly when working to windward; the loosely-woven flax sail-cloth of the age did not lend itself to the production of an efficient fore-and-aft rig which only became fully accepted in the second half of the nineteenth century. Their picturesque names, hoy, billander, brigantine and ketch, usually refer, however, to the construction of their hull and it was only towards the end of the next century that the specific rig of a vessel was employed as a primary factor in official records, perhaps indicating a change in the emphasis from a documentation assisting an intending merchant, to one providing a method of easy identification by officialdom both afloat and ashore. This rather hazy nomenclature applied, to some extent, to the larger vessels as well, for although by 1740 their size had reached, on average, some 300 tons burthen, they too were referred to, somewhat obscurely to our twentieth century ears, as 'pinks', 'hag boats', 'barks' or 'cats', with such intriguing sub-divisions as a 'square-sterned cat'. The particular name usually defined the shape of the stern or bow, in the case of pinks and hag boats a form clearly derived from the Dutch *fluits* and probably cheaper to construct, but less prestigious than the square stern, associated as it was with the lordly East India men, complete with a gallery of windows giving light and status to the main cabin. This form must, however, have been uncommonly vulnerable to a following sea or a billander's long bowsprit in a crowded dock but, nevertheless, became more common as the century progressed.

The bigger colliers were usually destined for the Thames; two hundred years ago not the deep uncluttered waterway of today and where a shelf above Woolwich gave only a minimum nine feet of water, but nevertheless a river capable of floating the deepest draught merchantmen to at least that point. On the outward voyage the Tyne bar also enforced limitations of size to colliers for the channel was only 40 yards wide and it was the custom to lay-up in winter when its dangers were at their worst. The tradition of a winter spent laid-up, with yards and topmasts sent down, did not entirely disappear until the arrival of steam towing to assist vessels leaving and entering the Northern ports (Chapter 10). In the seventeen eighties shipmasters were still rightly cautious of the dangerous north-easterly gales, the dominant wind of the area which made a dangerous lee-shore of the whole coast. It could embay an entire fleet amongst the sands of the Wash, known picturesquely as the "Devil's Throat". In 1783–4 between four and five times as many coal ships left Sunderland, outward bound, in the summer, compared with those leaving in the dangerous months January–March. This was a tendency away from an almost total winter stoppage in the early years of the century, an improvement no doubt stimulated by the modest improvement in

Builder's model, used instead of drawn lines, of the collier bark *Liberty and Property* 1752. The similarity to the hull form of the Dutch fluits is noticeable. *Science Museum, London*

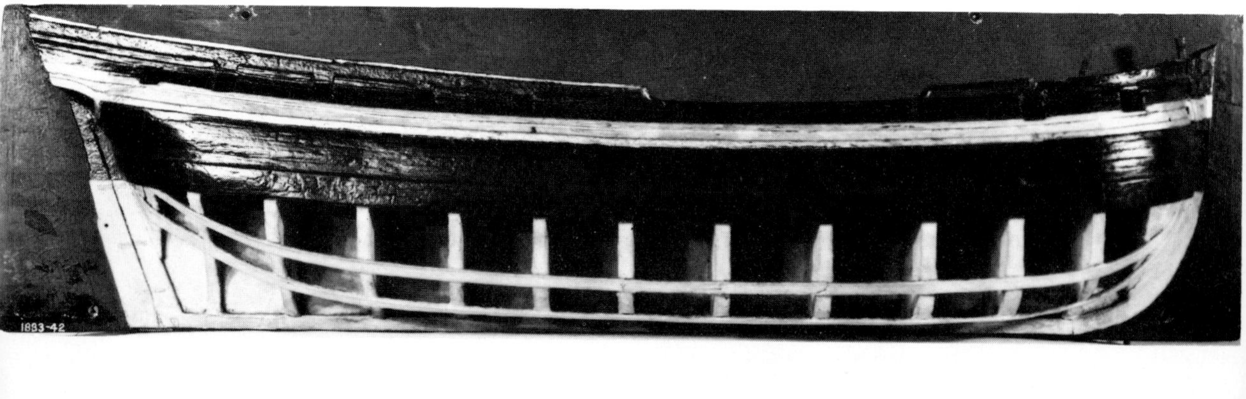

navigational aids (Chapter 10) and refinements in rigging and hull design which produced rather more weatherly ships. Insurance for both ships and cargoes was, of course, more difficult and expensive, if not unobtainable, during the time of storms and darkness. Sailing ships of this period could usually count on having to spend at least one month of the year out of trade anyway, refitting and repairing. At this time they were likely to lose as little profit as at any other season. There were also attempts by the coal-owners, ever keen to improve trade by restricting it, to place a definite date in the spring before which coal-loading was forbidden.

The size of the colliers reached an average loading capacity of 30 keels, for this tonnage allowed a nicely balanced compromise between the complex factors which controlled the size of ships during the days of sail and wooden construction. It was found by owners that the charges upon a smaller vessel were proportionately greater, particularly for crews, who were not only paid by the master, but fed. Alternatively, a large hull in a trade where calamity, always imminent, whether from enemies of the nation afloat, or sea, sand or structural failure, were ever at the commander's elbow, the loss of the ship would incur a crippling blow to the owners and was unpopular with insurers. The colliers had the further restrictions imposed upon them, of grounding in harbours which dried out, and of loading from keels, from whence the coals were thrown up to their decks which necessitated a low freeboard. Shipwrights had found that the construction of a wooden vessel within the limitations of the available technology and materials, any larger than about 150 feet brought immense problems aggravated beyond a reasonably economic solution, particularly when the hull had to sustain frequent uneven stresses and strains as it lay on uneven berths, loaded with a bulk cargo.

In a similar way the rig was developed to produce, within the restrictions of easily replaceable materials, an effective propellent which could be handled by the smallest crew. The apparent complexity of the cats-cradle of rigging aloft, almost beyond comprehension to our modern eyes, had to the initiated a clear logic of its own. It was adaptable to a wide range of weather conditions and, more important, might be repaired and maintained with a minimum of tools by a craftsman who had learned the grammar of his trade from long apprenticeship. Replacements for spars, cordage and repair to sails could, if not actually carried on board, be found almost everywhere a collier-bark might find herself. Damage to the hull admittedly presented problems which could only be solved with specialist assistance and equipment. However a hull capable of loading 300 tons might be careened, without the elaborate machinery of a slipway or floating deck. The hold was emptied and the hull strengthened internally with temporarily fitted diagonal strutting athwartships. Multiple purchases, from the shipyard or adaptation of the vessel's own gear, were attached to the mast-heads and anchored to the quay, while spars lashed between the quay and the vessel allowed the hull to cant-over, until the keel was clear of the water, when the purchases were

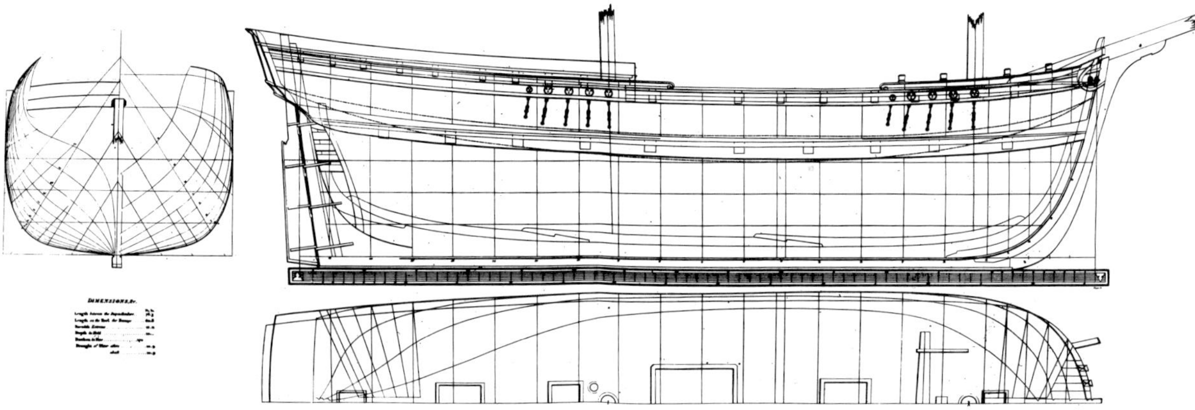

Lines of a Collier Brig, 170 tons, from Steel's *Naval Architecture*, Plate 22, published 1804.

shortened. When one bilge of the wooden hull had received attention the vessel was brought upright, swung round and the process repeated. Less spectacularly but more regularly, colliers were 'breamed'. Allowed to dry-out on a hard shore, fires made from furze cut by the crew were lit to heat cauldrons of tar, or merely to provide burning brands of reeds with which the marine growth which inevitably accumulated below the water-line, could be removed and the liquified tar re-distributed.

Such expediencies took time, patience and ingenuity, long hours knee-deep in water and mud and the ever-present danger of crushed hands or worse if gear parted without warning. But the capital equipment needed was minimal. The work could be done in widely differing places along the river shores of Eastern England known by tradition as providing natural advantages and there was beer to celebrate success. The carpenter, who was to be found on the crew lists of most collier-barks during the eighteenth century, certainly deserved his modest £2.15s. per month. Besides supervising with the mate such operations as careening, he would shape spars from baulks of timber, search out and at least temporarily effect a cure for leaks, repair the vessel's boats (always susceptible to damage in crowded harbours and tideways), and generally spike, strap, caulk and cosset his vessel so that it might make port in safety.

Of course, there could be distortions to the evolution of a vessel's rig and hull form arising from purely material factors. The provision of bounties, which were paid if extra apprentices were carried, could inflate crew-lists as did the necessity of carrying extra hands to act, in emergency, as gunners. These were employed working unnecessarily elaborate sail-plans. The trade could be sheltered from competition and the resulting stimulus to adopt technical innovations already in use in foreign vessels by an overt national policy or less explicitly, but equally effectively, by private arrangement between producers and shippers. The archaic discharging limitations maintained in order to support a series of overlapping sectional interests kept capital invested in shipping unprofitable and literally tied up for weeks, and did little towards stimulating the shipbuilders of colliers to evolve faster, more weatherly vessels. The time spent at sea was the least pressing problem; it was the inordinately long wait in the Tyne and the Thames which called for reduction and regulation. Even amongst its contemporaries the collier fleet earned for itself a name for conservatism, and this in a century not noted for rapid technical advance. The long delays to which they were subjected might well have contributed to this as seamen and apprentices could always be employed overhauling and maintaining rigging and gear so that its basic inefficiencies and over-elaboration were disguised and alternatives were never explored.

The shipways from which these colliers came were the North-East harbours of the coast rather than the ship-yards of Newcastle which did not launch a ship of more than 30 keels before 1750 when the *Russell* was constructed at Headlam's yard. An unwelcome result of naval success during the Anglo-Dutch wars of the second half of the seventeenth century, had been the introduction of cheap prizes of war. Under their impact upon the market the ship-yards of the South East had succumbed. Never a flexible form of enterprise, ship-building was particularly susceptible to competition and market fluctuations. In the 1720s as Defoe explains "Dutch flyboats (fluits) taken in war, thrust themselves into the coal-trade for the interest of their capture . . . These Dutch vessels costing nothing but the caption, were bought cheap, carried great burthens . . . building fell off for want of price and so the trade decayed". We have other evidence besides that of the itinerant Defoe for, by this date we can assess the relative prosperity and vitality of various ports from tonnage returns and Whitby, followed by Scarborough and Sunderland, all established themselves at this time. North and South Shields had to free themselves from the monopoly which Newcastle's freemen claimed and enforced in this, and much else, before they could equal their neighbour's efforts higher up the Tyne.

Whitby today, still a place of great individuality and haunted by the ghosts of its former fleets, was a port without indigenous wealth, whose only resource was a natural harbour. Set on the edge of the moorland upon "a little nameless river, scarcely indeed worth a name" as Defoe quite incorrectly supposed, only the alum workings along this stretch of the Yorkshire coast provided a limited local wealth. Beginning in the seventeenth century this had necessitated an embryo fleet of colliers to import the disproportionately large quantities of coal needed. There had been piers at Whitby in Tudor times when Henry VIII ordered their maintenance at royal charge and the timber was supplied from the "King's Woods". The

Tyne, with its bar at its entrance, did not attract ship-builders to the same extent, perhaps because at this time its energies and capital were fully absorbed by the coal-pits and their ancillary industries. There are many parallels to be drawn from the days of sail of a similar situation where a small port, denied natural resources or the position to be the outlet for a wealthy hinterland, served the needs of a more happily provided neighbour. In 1710 there were said to be 120 vessels of 350 tons and upwards, mostly colliers, owned there, a growth stimulated by the fact that eight years previously an Act of Parliament provided for a substantial improvement of the piers begun in Tudor times. These were financed by coal dues at the rate of ¼d. per chaldon on coal shipped from the North East ports and in one year no less than £4,000 was raised in this manner; not a surprisingly large amount when one learns the size of a fleet of collier-barks. Congregating in Harwich harbour because their captains had refused to sail until certain grievances were met, the accumulated host numbered between an astonishing 600 and 700. To serve this fleet another Act in 1740 ensured that Whitby harbour was further improved and deepened, which meant that larger ships could be built there than hitherto.

As prosperity increased more banks established themselves in Whitby's Church Street, Grape Lane and Baxtergate to finance and underwrite maritime ventures while Charlton's Mathematical School furnished instruction in "Arithmetick, Algebra, Navigation, Guaging, use of Globes and all other Mathematical Instruments" for the benefit of aspiring apprentices and mates. Messrs. Fishburn and Broderick, who built three of Captain Cook's vessels, had taken over a ship-yard from Jarvis Coates, who had started a similar undertaking in 1749; while G. & N. Longborne, who launched the explorer's other ship, had succeeded Richard Simpson.

One should not fail to remark, at least in passing, how the skill and enterprise of Whitby men and builders found further expression in the whale-fishery. The versatility of the eighteenth century vessel is well illustrated by the following typical advertisement of a contemporary Whitby vessel, even if one allows for the normal exuberance of a copy-writer. It states that "The good ship *Advice,* square-stern, Whitby built, 207 tons measurement, will carry 17 keels of coals, is calculated for the Baltic, Greenland, (whale-fishery) coal or coasting trade on a light draught of water, takes the ground well, shifts without ballast, sails remarkably fast, well found with all kinds of stores, and may be sent to sea at an easy expense, had a thorough repair last winter at Whitby . . . ". Many whalers made voyages in the coal-trade between whaling in the Arctic which was only undertaken between May and October.

No eighteenth century vessels could sail without ballast below decks. But their ability to move from one berth to another with a swept hold in port, by warping or towing by row-boats, was an essential quality if undue expense and trouble was to be avoided by the master. This was always mentioned in advertisements.

South of Whitby, nearby Scarborough, standing on its salient of sandstone, now lacks some of the authentic salty tang of Whitby, but was at one time the leading collier-owning port of England. Rising from the status of a small and far from elegant fishing village it also rebuilt its harbour from the proceeds levied upon the buoyant coal-trade. As the prosperity of Scarborough faded by the mid-eighteenth century Sunderland took its place as a ship-owning port, having the added advantage denied to its southern competitors of commanding the entrance to a river flowing through a county rich in coal-measures. Both ship-building and coal had been linked at Sunderland since the Middle Ages and the depth of the harbour was improved in 1717; within two decades Newcastle and Sunderland between them shipped a million tons of coal coastwise, largely to London. Besides this trade down the North Sea, a greater proportion of Sunderland's colliers traded to foreign ports. There was an important export trade to

Colliers off Harwich by F. Swaine 1775, similar to the barks which were taken into the navy and used by Cook and Flinders. *National Maritime Museum, London*

Holland, although the frugal Dutch housewife was said to prefer 'splint-coal' from the Firth of Forth as this was dust-free: and indeed it was frequently pointed out that, because it was not taxed, Newcastle coal was cheaper in Amsterdam than in the Pool of London. The colliers then sailed from Holland to Norway in ballast, to stagger home with a cargo of timber stacked high on deck for the North East ship-yards and the building trade.

Sunderland had suffered long from the attempts of the Hostmen of Newcastle, who effectively controlled the policies of their city, to stifle its development. In 1661 the Hostmen procured an impost of one shilling per chaldron on all coals exported from Sunderland. Later in the eighteenth century Sunderland retaliated to this and other attempts at discrimination by exempting Yarmouth colliers from harbour dues and so tempting them to load along the Wear rather than at Tyneside staithes. The Tyne then refused to load vessels which had taken cargoes from the Wear. It was fortunate for the consumers that such competition to Newcastle existed, based as it was upon Sunderland's access to virgin seams of good quality coal in the immediate hinterland of the port and the absence of Hostmen deeply entrenched in a powerful corporation. It provided at least a measure of protection to the London consumers from the cartel founded by pit owners in 1710. Known at the beginning of the eighteenth century as the "Regulation", it sought to create a monopoly and limit production by restriction and apportionment of output to individual mines and designation of their shipments to London. The concentrated market in London and the coastal ports together with the monopoly in supply held by the Northern coal-owners, presented continual temptation to form restrictive arrangements. There was some excuse in earlier days, when the Northern coalfields gained the name of "Britain's Peru" and attracted southern capital of a 'high risk' nature which could cause commercial dislocation both by its rapid investment and equally rapid withdrawal. At one time the "Regulation" laid down the general rule that the Tyne pits should divide the available market with the Wear in the proportion of three to two. The "Grand Alliance", as the cartel was later known, worked not only to regulate the market but tried, more excusably, to secure way-leave for the growing lengths of wagon-ways and to act as a pressure group in London to put the owners' point of view. As the older pits, close to the Tyne, were worked-out by the seventeen hundreds, coal-measures further north and west were sunk at considerable cost and exploited some distance from the river-side staithes and the extended wagon-ways placed local land-owners in a commanding position to extract the highest possible tolls. By combination the coal-owners could bring greater pressure to bear on land-owners so that tolls were realistically negotiated. If a ship-owner loaded a cargo outside the controlling group of pits the long arm of the Regulation reached out and the London agents of the owners could prevent lighters being provided for the colliers when they arrived in the Pool.

The effect upon the vessels in the coal-trade was to engender endless frustration. Masters were kept at anchor in the Tyne if a colliery had exceeded its monthly quota of production before the end of the allotted period. Whitby ships tended to be worse off as they had far fewer coal-owners with a direct interest in their profitability than the Newcastle and Sunderland colliers and might be kept waiting eight or ten weeks for cargoes while the trading season slipped by. Their crews were paid by the voyage and so did not present a continuing expense as the waiting dragged on, but the engagement of a satisfactory ship's company, particularly if the press-gang had been recruiting, was not the matter of a moment's work. Apprentices, of course, remained aboard to be fed.

The ship-masters did their best, in imitation of the coal-owners, to organise themselves into a pressure-group, complaining bitterly about 'interlopers', that is to say ships diverted from other trades to take advantage of a temporary quickening of demand. But their numbers were too numerous, too widely spread and too easily subjected to pressure to organise effectively. The 'crimp' of the seventeenth century Roomlands, near Billingsgate, who supplied dockside labour and acted as buyer and go-between, and with whom the master bargained on the Thames was now the factor of social standing and wide commercial connections. The lightermen acted as crimps as well and could effectively 'black-ball' a ship whose master was unco-operative, with a resulting delay to unloading.

In 1769 the establishment of a covered market in London, sponsored by the Corporation of the City, with an exclusive membership of agents and factors formed into a Society, did little to improve the situation from the ship-master's point of view. It replaced the old informality of a ground adjacent to Billingsgate and the easy-going contacts that could be initiated in nearby coffee houses. Like the ownership of the colliers, the new Coal Exchange was divided into 64 shares, distributed amongst the newly-formed Society of Coal Factors. The City of London stipulated that no one could import coal without its sale passing through the Exchange where Factors alone were authorised to sell and where cargoes were sold by 'wholes', 'halves' or 'quarters'. The Factors, in order to be able to judge the market tendencies kept a man at Shields who communicated regularly with the London office. He advised of production prospects, details of shipping and weather conditions as it affected arrivals and departures over the Tyne bar. A spell of North Easterly gales could pen-up loaded ships; the appearance of determined privateers off the coast, a 'hot press' to recruit seamen in war time, all had their repercussions upon the market. Individual Factors had shares in northern coal-mines, or at least relatives who held them and they had interests in colliers and pressed the sale of their own cargoes as against those of others and some authorities have considered them the southern outpost of the Regulation. The position of Factor was to become a springboard for wealth and status in the London and Tyneside financial communities while there was a natural tendency

Margate harbour in the 1850s. Even after the coming of the railway Margate was using seaborne coals for brewing and gas-production. *Margate Public Library*

for them to branch out into shipowning and insurance. Their training in efforts to avoid critical fluctuations of price, to parry parliamentary enquiry and shield the distributive system they controlled from mounting popular criticism, must have provided an admirable training for such a career.

In 1728 the master-owners of the colliers had tried to protect their own interests and with a commendable self-discipline set up an office at the *Dog Inn* in the City where a register of inward cargoes was kept. This recorded each arrival and on completion of unloading a certificate was issued; it was arranged that without this, another cargo could not be secured from the Tyneside fitters. It attempted to ensure an orderly reception of cargoes with no especially favoured vessel discharging out of turn. Needless to say, it was hotly opposed by the Factors and finally had to be abandoned.

Sunderland reached a position at this period when it was shipping nearly half the tonnage of its rival, double its own output of twenty years before; but it never overtook the Tyne. Both joined, temporarily, in complaining against common problems; supposedly greedy ship-masters and London factors ready to strip them of hard-won profits earned from a precarious venture. But it was an entente which evaporated at the first sign of proposed joint action and the sacrifice to profits that such combination might entail. Sunderland, unlike the other ship-building centres of the Yorkshire coast, did not falter and die with the passing of wooden ship-building. Its later development is for another chapter, but the price it paid in slums and industrial dereliction as it in turn was superseded has only recently been ameliorated. Between the Wear to the north and the older ports lies the Tees and its broad estuary. During the eighteenth century when others seized their opportunity, this area which was later to develop into an important industrial complex of steel, steam and grime remained untouched by the effects of the coal trade. The banks of the Tees, despite trade reaching as far inland as Yarm, had to await the building of the iron road in the next century which linked it with the southern limit of the Durham coal-fields, before it could emerge from what was in effect a pre-industrial economy.

By the late eighteenth century the collier-seamen and their dependants were an overwhelmingly dominant section of the community at North and South Shields while, to a lesser extent, they lent character and colour to the ports of Blyth and Sunderland. These were coastal sailor-towns whose population looked to the sea and to a large extent a single staple cargo for their livelihood and profit. Unlike the men and boys who composed the crews of the ocean-traders the complements of the individual collier ships were frequently drawn together, either on shore or afloat. In earlier days this naturally occurred when to lay up vessels during the winter months was taken as a matter of course in the trade. Subsequently as this tradition died, the vagaries of the delivery of freights at the shutes, something implicit in the restrictions of the coal-owners 'Vend', or flooded pits, foul winds or gales all conspired to accumulate fleets and crews together. These gatherings provided opportunities, rarely accorded to the crews of ocean-voyaging ships, for men to exchange at length the details of their grievances which lost nothing in their telling from the crew's boredom through forced inactivity. On the Tyne or the London River they gathered in fo'c'sle or riverside drinking houses, a self-aware section of the seafaring community, closely knit by a common calling and dialect, celebrated in song, inured to hardship beyond even the standards of their time and holding the conviction, reinforced by comment in Press and Parliament, that they held a vital place in the nation's life. This gave them an opportunity and a motivation, denied to other workers, to take counsel and evolve forms of industrial action to remedy their grievances. The turbulent North Country at all levels of its varied society gave them examples to follow. The keelmen, who like the seamen, had the advantage of inhabiting the traditional settlements of river-workers at Whickham, Dunston and especially Sandgate, outside the old walls of Newcastle, staged 'mutinies' as their strikes were called by the authorities. No doubt the magistrates had in mind the traditional punishment for such action when referring to them in this way. These disruptions were a feature of Tyneside life and their first, in the fifteenth century, has been quoted as the earliest recorded example of a strike by a specific group of workers. Pitmen and hewers were still craftsmen with a tradition of independence, men made bold by a daily contact with hidden and half-understood dangers. All were very different to the other dissident groups of the time, the round-shouldered weavers, ailing glass-workers, undernourished agricultural labourers, dragging out an existence in adjacent regions.

The ship-owners of the North East combined to strengthen their position, collier-owners combined to protect their interests and maintain the 'Vend' and keelmen joined to press the fitters for a better bargain; with these examples before them why should not collier-seamen do the same to establish adequate manning scales and wages? The sailors, when they brought pressure to bear faced the weapons of blackleg imported labour and in 1824 the owners brought in sailors from as far away as the Orkney and Shetland Islands. The freighting of vessels from outside the tightly-knit area of the coal-ports was rarely resorted to for it discomforted the ship-owners even if it assisted the Lords of Coal. Indeed the ship-owners were in the most vulnerable position. They were too numerous for effective informally agreed arrangements, such as those by the coal-owners of the eighteenth century, while by tradition and inclination they declined to form the forceful mass-movement, with its lively solidarity, which came naturally to the seamen. The victimisation of the union leadership was encouraged by the captain-owners through the mutual insurance clubs which provided some focus for their interest. Through them names of seamen were circulated whom it was considered inadvisable for skippers to sign-on. Unlike the colliers ashore, who encountered their own version of these problems, the shadow of impressment for service in the Navy, or the withdrawal of exemption, at best a tenuously held privilege, was always at hand to be used as a deterrent against the seamen. The press-gang was a unique threat which hung over the collier-seaman and his family and was employed by the civil authorities to assist in strike-breaking as late as 1822. But the presence of the King's cutter anchored off Shields to transport a seaman to indefinite years of exile under the rule of the lash did surprisingly little to cow the Geordie sailors.

Concerted action is first recorded in 1768 when four or five hundred seamen of North Shields marched to Sunderland with colours flying and having formally voiced their grievances, prevented any vessel from leaving port until they were met. From then on there were intermittent disturbances until the mid-nineteenth century when the fundamental structure of the industry changed. However, seamen were not always in direct conflict with ship-owners. Both groups joined forces in 1775 to withstand pressure from the coal-owners who insisted that shippers should pay an increased price for coal at the staithes amounting to an extra shilling per chaldron. The crews, although only normally signed-on, and therefore victualled, to a vessel loading at the shutes so that they were available to assist in trimming and warping, were enlisted by the owners and fed, providing they agreed to dissuade recruitment to non-union ships. The seamen actively prevented those ship-owners who had agreed to the increased price demanded by the coal-owners from proceeding to sea by towing back with longboats any vessels who attempted to venture out over the Tyne Bar. To further strengthen their argument a boom of keels and wherries was improvised to blockade the way to sea should any other ships be rash enough to sail; an expedient which was again resorted to during a grimmer phase of the seamen's struggle in 1815.

The strike of 1815 was the only occasion when loss of life through militant action is recorded and it took place at a time when wide-spread unemployment had embittered relationships. This strike had lasted, with commendable good order, from the beginning of September until the end of October. Eventually the Royal Navy, who to its credit deeply resented having to participate in a situation which might lead to their clashing with merchant seamen, was asked by the Civil Authorities to take control and re-open the harbour. The seamens' grievances concerned the undermanning of ships and the signing on of boys to serve before the mast where previously men were only considered adequate; both practices which had gained a reluctant acceptance under war-time conditions. The strikers anticipated a modern expedient of placing their opinions and aspirations before the public by buying advertising space in the radical *Tyne Mercury* and distributing broadsheets, in the manner of the ship-owners. But despite the use of such civilised exchanges between the contending groups violence eventually erupted and in a riverside confrontation several seamen died from shots fired by marines and troops. The Newcastle magistrates, with their roots in the local community, had shown a commendable degree of appreciation of the seamen's grievances and deplored the recalcitrance of the ship-owners; but they eventually succumbed to the pressures from Whitehall, ever apprehensive of a local breakdown of order spreading to endanger the whole establishment.

The solidarity amongst the membership of the seamen's union was reinforced with a surprising element of self-imposed discipline. Robust treatment was undoubtedly meted out to press-gang informers who betrayed collier-seamen to the recruiting officer and to blacklegs, but they could be equally severe to their own members who, by drunkenness or disorder, might bring dishonour upon their cause. The union seamen's patriotism was never in doubt. In 1792, during a strike, the newly built H.M. sloop *Martin* was sent to the Tyne to strengthen the power of the civil authorities, but instead went ashore on the Herd sands. The seamen, whom the naval vessel had been despatched to intimidate, launched boats and assisted in refloating the stranded vessel. No doubt the opportunity to demonstrate their professional skill, as well as loyalty, to one who had been sent to overawe them, added to their enjoyment of the piquancy of the situation.

The centres of emergent union activity were stimulated by a response to the fact that the men who served afloat aboard the ships and the craftsmen who built the colliers, could look forward to only an inadequate patchwork pattern of of assistance should ill-health, shipwreck or old-age bring them to penury. The paternalistic charity of the nineteenth century was yet to develop, but its twin demand for 'self-help' stimulated the foundation of friendly societies by and for seamen. Besides providing relief and a modest status of respectability for their members, they inevitably became a cover-activity for unionism, either as a

A group of vessels in the harbour at Whitby which recalls the shape of ships in the eighteenth century. The *Lively*, extreme left, was built 1786 and lost 1888. On the right is the *Alert*, also built at Whitby, launched in 1802 by G. and N. Langborne who built Cook's *Discovery*, ex *Diligence*, used on his third voyage. Both the *Lively* and *Alert* were built as sloops and later lengthened and converted to schooners.
Sutcliffe Gallery

natural extension of their objectives or to circumvent the out-lawing of trade-unions. Clandestine unions flourished, serving the seamen as friendly societies for the first three decades of the nineteenth century, although not every society was involved in this way. One which was, the Loyal Standard Association of South Shields, and whose membership was restricted to men who had served an apprenticeship at sea, later had branches in Bridlington, Scarborough and Whitby was formed as a union in 1824, but reformed as a friendly society in 1832. The British Union Society and the Seamen's Amicable Society all served the needs of Tyne seamen while the Master Mariners' Social Society, with a nicety of choice in its title to distinguish its different area of recruitment, catered for the skippers. The South Shields Shipwrights Association was perhaps the first union for shipwrights in the country and dated its foundation from 1795. Closely followed by the Association of Shipwrights in 1798 both provided widows' benefits, relief in time of sickness and the assurance of the member's burial was accorded a suitable dignity and decorum that in life the recipients had all too often lacked. The South Shields Shipwrights Union vied with the seamen in militancy, but it was essentially a conservative craft-based union although it was sufficiently prosperous, and courageous, to finance a ship-building project for its unemployed members in 1826 and launched a 400 ton ship.

The existence of these and similar societies provided an admirable experience in self-governing and responsible organisation while giving their members an opportunity for the sober display of solidarity and status in their annual parades. But the competition which inevitably grew up between the numerous societies, in both of their aspects, weakened the organisations and led to financial problems. In the same way as the launching of the ships they built were recorded the shipwrights' associations were delightfully celebrated in Newcastle and Sunderland lustre-ware jugs, punchbowls and mugs which were manufactured from 1780 until well on into the nineteenth century.

The efforts of an earlier age had done something to provide for the seaman should he, by chance, survive the end of his working life. The Trinity Houses of London, Hull and Newcastle maintained charities for those who 'having served long and honestly' could find votes from worthies to support their application to the alms houses the ancient organisations administered. The foundation of the Seamen's Hospital at Greenwich in 1821, expanded this work by providing medical treatment when it was initiated by a group of philanthropists, led by William Wilberforce. Established aboard the hulk *Grampus,* an old man-of-war, in the Thames, its conditions of hygiene would have appalled a modern physician. Nevertheless, the floating hospital admitted any seaman, regardless of creed or colour. The provisions, recorded in 1821, indicate that an inmate should be provided with "14 oz. bread each day, ½ lb. butter each per week, 6 oz. cheese each per week, meat 3 days a week – ½ lb. each when dressed with vegetables.

Gruel for breakfast. Three pints of beer each per day". It must have seemed a haven of plenty for a broken seaman, weak from illness, taken out of the gloomy fo'c'sle of a collier brig at anchor in Woolwich Reach to a bed on the 'tween decks of the old man-o-war.

At Whitby, as early as 1675, a fund had been raised and maintained by the shipowners, masters and burgesses of the port. It stipulated that any shipowner, master or sailor of any vessel above twenty tons should pay a toll; the owner contributing a shilling, the master sixpence and the sailor twopence if his wage exceeded twenty shillings, a penny if it exceeded ten shillings; the contributions to be made upon every voyage past Whitby. At Lynn ships which were 'foreign', that is to say were not owned at the port, or in cases when the buyer and seller of the cargo were not Freemen of the town, paid eightpence per chaldron for the relief of the seamen's dependants. These and other local expedients were supported, in theory at least, by a 1747 Act of Parliament which set up a general fund produced from compulsory deductions, extracted at the Customs House from the seamen's wages. Intended "for the Relief and Support of maimed and disabled Seamen and their Wives and Children", the fund was rarely administered with the same high ideals that saw its initiation. The fund's mismanagement became the cause of common complaint in later years amongst those it purported to help. As a collier seaman told Henry Mayhew in the eighteen fifties, "We pay to the Merchant Seaman's Fund 1s a month but I never met a man who knew what became of it". The degree of organisation that was required to collect signatures to support petitions to Parliament in an attempt to remedy the situation, assisted in strengthening a growing militant unionism.

The close geographical proximity of the coal ports made concerted action between groups of seamen relatively easy to organise. By the eighteen fifties they were able to organise support in Lynn and Ipswich to remedy grievances of under-manning and low wages. Occasionally, as in the eighteenth century the apparent interest of seamen and ship owners coincided. In the late eighteen forties both parties organised meetings, petitions and demonstrations to oppose the repeal of the Navigation Laws which, for four hundred years at least, had protected British shipping and ship-builders. When, despite all their efforts the repeal received the assent of Parliament in 1849, the yards of every British vessel in port on the East Coast were crossed in mourning and the foreign ships' masts were dressed over-all with bunting. The objectors did, nevertheless, by their efforts, gain some reservations upon coastwise shipping for a further four years.

7 A Collier's Nag

"If his breeding has been North of Yarmouth, he is distinguished with the title of Collier's nag; and indeed he is a rare horse that will never fail you in bad Weather being as insensible to Rain, Cold or Thunder as a Cannon-bullet."

The Wooden World. E.Ward 1701.

AUTHENTIC accounts of a working life, written from personal experience, rarely survive from any age. We are therefore particularly fortunate to have in the autobiography of William Richardson a convincingly authentic and varied picture, painted with detail and colour, of a life afloat in the latter eighteenth and early nineteenth century. It is moreover an account cheerfully told of a life spent in the fo'c'sle and on the weather yard-arm, not on the quarter deck. William served his time in a wide variety of merchant men as well as a slaver and was twice pressed for service in the Royal Navy. But here we are concerned with his boyhood experiences in the North East coal trade and the light it throws upon the men and ships of his period.

Born at smoky South Shields in 1768, William Richardson was the son of a seaman who, like many of his countrymen, had emigrated to Tyneside from Scotland, to become one of a large group of expatriate sailors and keelmen. Throughout the narrative of William's years serving in the colliers there recurs a theme of strong family ties reinforcing a feeling of local community. It was a community which existed ashore in the half dozen coal ports and afloat, aboard the great coal fleet. It was a self-aware group of seamen where professional mis-judgement was tolerated, but never accepted, folly laughed at, but callousness rarely passed over and the refusal to render help when it was needed, never forgiven. The hardships and dangers of the life at sea were, at least in part, made tolerable by the working relationships formed aboard ships and strengthened by long-standing links of family and fellowship. More widely, this attitude permeated the whole fleet whether it was at sea, in port, or lying in the crowded anchorages. Here, in situations where the problems of manoeuvring and working the unwieldy ships, with wind and tide as the only motive power called upon the ingenuity and co-operation of the ships' crews and the skill of their commanders which were tested to the limit.

As the captain and part owner of a bark, William's father was in the position to send his son to a modest boarding school where he was introduced to the study of mathematics. This was extended later by a schoolmaster at Shields who "had taught most captains of the place navigation". At the age of twelve Richardson was considered old enough to ship aboard the parental command *Ravensworth,* built in 1765 at Newcastle and of 196 tons. She was bound for the Pool of London with coal and the authenticity of the account is reinforced rather than diminished by his frank admission of seasickness and a longing to be at home in Shields whose identifying pall of smoke from the salt-pans could be seen far out to sea. Young William had not long to wait before he had the first of his numerous experiences of war at sea. Upon the return voyage to the Tyne, down to the North from London, they had reached Flamborough Head, a favourite point for ambush, when they encountered a French privateer. It was a cutter crowded with armed men, which gave them mounting cause for concern. The collier *Ravensworth,* like the great majority of eighteenth century merchantmen carried carriage-guns, some mounted dummy wooden ones, known with a nice sense of humour as 'Quakers', and she "had six guns, and four of them we could fight on one side." Preparations for a fight were made. The braces, used to control the yards, were led through the gratings on deck, to be worked from the relative safety of the lower deck while a short iron tiller was shipped on the rudder-post where it passed through the after-cabin. There were loopholes cut in the hatch coamings through which small-arms' fire could be directed by defenders standing on the lower-deck to deter boarders, but the collier's crew were not called upon to prove their determination. For whether the Frenchman did not consider the capture of a mere ballast-loaded collier with such a well-prepared defence worthwhile, or whether he was suspicious of what was, in fact, only a fleet of colliers coming up astern, in the event he crossed the bows of the *Ravensworth* and made off.

They were perhaps more than usually fortunate for the daring of the privateers was notorious. A few years before, a French privateer, standing in too close in hazy weather, ran on the reef at Saltwick, about a mile from Whitby. Undetected, despite their precarious situation, they were so confident, or else intended to prove they were, that they sent a local fisherboy into Whitby to fetch them an anker of beer. Suitably refreshed, they hauled off on the rising tide to safety. Another free-lance man of war, short of supplies, landed a party on the cliffs between Scarborough and Whitby and made off with thirty sheep and lambs.

Rare contemporary model of an eighteenth century collier cat or bark, showing the absence of a projecting head, a tradition which lasted until the last days of the Geordie colliers, and 'pink stern'. It was in a vessel such as this that Richardson first went to sea. Scale—one twentyfourth, the vessel was 98 ft, beam 28.6 ft.
National Maritime Museum, London

The telescope of the *Ravensworth's* commander and those of his fellow captains must often have been directed towards the flagstaff upon the south cliff of Whitby. Here an ensign was hoisted if a privateer was in sight to the south, a pennant if the Frenchman was to the north while a pennant above the ensign was that the enemy was directly out to sea.

In 1757 three French privateers, two snows mounting 16 guns and the third a lugger, chased a fleet of 51 colliers into Whitby harbour, while in the same year a Dunkirker sailed into her home port with ransomers representing 15 ships, captured in the North Sea.

The manner in which the eighteenth century privateers plied their trade was, on the whole, surprisingly civilized, even if their benevolence was ultimately dictated by hard economics. Captured vessels with bulk cargoes, and the colliers were, of course, numbered amongst these, unless their freight was in demand in the privateer's country, usually ransomed at sea for about a quarter of their total value. Especially valuable, portable freight was trans-shipped to the aggressor, if conditions at sea permitted, or a secluded anchorage could be found. In either case a hostage was taken, usually the first mate. By the early eighteenth century the system had sufficiently advanced for printed forms to have been used by the French privateers who expected the captain of the capture to sign as an undertaking to pay a ransom. The merchant-vessel, when released to proceed was then, very sportingly, entitled to be granted twenty-four hours immunity from further attention by a French privateer. The ransom money was ultimately negotiated through neutrals, as an example of which we may quote the *Charming Molly,* collier bound for Amsterdam in 1780 and taken by the French who ransomed her for 400 guineas in gold. In this way a privateering enterprise could pay handsome dividends to its investors without flooding their home ports with captured tonnage and a resulting crippling drop in the value of shipping, while the problem of shepherding back slow-sailing captures to harbour was avoided.

The adventures that might be expected in war-time by the coal-fleet are exemplified by the adventures of the *Ranger.* In 1798 on passage to London, coal laden, she was captured off Huntly Fort by a French privateer, of six two-pounder guns. Four other vessels were taken on the same day and sent to Norway. The master of the *Ranger* and two other Englishmen were turned adrift to manage as best they might in an open boat, but were rescued by the *Betsey and Molly,* of South Shields. However, the incident was satisfactorily concluded, as the *Norwich Gazette* reported, "the privateer has been captured two of the vessels are retaken (one of which is said to be the *Ranger)* by one of his Majesty's armed vessels, are sent into Shields. The same privateer captured off Whitby, the *Dispatch,* from Shields to London with coals and glass. The privateer is a new vessel, schooner rigged, with two top-sail-yards, has a black bottom and a small woman's head".

She must have made short work of the slow, heavily laden colliers and was typical of the privateers sent out to find a quick way to fortune and only vulnerable to an equally fast-sailing man-of-war.

The *Ravensworth* avoided the attention of the privateers and after three successive voyages under the parental eye, during which William exchanged the comparative luxury of a bunk in a cabin aft for a hammock slung in the fo'c'sle with the other ship's boys, he was formally apprenticed. His apprenticeship was for seven years and his indentures bound him to the captain of the ship *Forester*, of 350 tons. At the end of his servitude he was entitled to an award of £25 and the freedom of Newcastle, for the commander of the *Forester* was a freeman of the port while his father was not. This status carried with it pecuniary advantages for it granted an exemption from Newcastle Townhouse dues of five pence per chaldron, liable to the shipper of coals.

The seventeen eighties still saw colliers put out of service in the winter months and while the *Forester* was laid-up William attended the Trinity School at Newcastle to improve his navigational skill. With the coming of spring the collier fleet bent their sails, rove their running rigging, tarred their shrouds and prepared for sea again. But the war against the United Provinces was still in progress and to the normal hazards of the North Sea were added enemy cruisers, privateers and the uncertain political situation which made it impossible for a vessel which was at sea for weeks at a time to ensure who was neutral, benevolently neutral, on the verge of war, enemy or ally. Convoying was in force and Richardson provides a detailed and telling account of the appalling delays and dangers which seaborne commerce suffered. The *Forester* crossed the Tyne bar, coal-loaded, and bound for Lübeck. The export of coals was booming during the later decades of the eighteenth century, principally to France, the Baltic, and even occasionally as ballast in barks outward bound to the Americas. After a refit it was usual to secure a freight destined for a distant port rather than the shorter haul to London, for sails and ropes had been newly overhauled, anchor cables were unworn and spars and hulls in good condition. But delay for the *Forester* was inevitable until a convoy escort was available and a great host of merchantmen had consequently accumulated. At last a frigate was seen "with a signal for the Baltic convoy flying at her masthead". The collier made a hurried departure and over-anxious to maintain her fast-sailing protector in view, carried away the main top-mast during the night, an eventuality that Richardson, with the traditional understatement of his profession, does not consider worthy of enlarging upon, but which must have engaged the crew in a major task of skilled seamanship and improvisation. Upon reaching the safe anchorage of Leith Roads they awaited the convoy to assemble once more, incurring a delay this time of six weeks. The only breaks in the monotony of the enforced wait were occasional rumours of a Dutch fleet threatening the ever-growing flotilla of relatively helpless merchantmen which made

A sailor-made model which survives from the time when Richardson was at sea. The *Brotherly Love*, built as a collier cat in 1764 survived for over a century. Despite the somewhat crude workmanship it gives an excellent idea of an eighteenth century hull.

Photograph John Lewis. Hull Maritime Museum

an increasingly tempting target as the numbers increased. Eventually four hundred vessels awaited the opportunity to sail and this was by no means an exceptional accumulation.

The arrival of a naval squadron, including ships-of-the-line, enabled the armada to weigh anchor and proceed to sea. Uneventfully the *Forester* crossed the North Sea and arrived at Elsinore. Elsinore, at the head of the Kattegat commanded the entrance to the Baltic and the *Forester's* captain took the opportunity, before clearing, to load stores of tea, gin and sugar which by this time, unless acquired from a smuggler, carried a crippling imposition of tax if purchased in Britain. Their freight of Newcastle coals had to be lightered out at Lübeck while the bark lay at anchor some way off the ancient port and they then took aboard ballast "brought off in boats each carrying a ton, the people throwing it up into the ship with their paddles". Young William does not hesitate to note, with a mixture of patronage and mild disgust, the outlandish and meagre diet of the natives; comments which were to find echoes in a hundred accounts by compatriots on the habits of those not fortunate enough to have been born British.

For the return voyage a valuable freight of iron, hemp and tallow, the traditional exports from the Baltic ports, was secured to be loaded at Cronstadt on the extreme eastern shore. It was a traditional part of the ship's company's contract to assist in loading the cargo. The labour was long, unpleasant and unremitting; for the longer a ship remained in port the more she accumulated harbour dues. Richardson refers particularly to the extended hours of daylight during summer in northern ports when only a belated sunset brought rest for the weary crew.

Their return from the Baltic was a stormy one, undertaken at the end of the season, testing the deeply loaded wooden hulls almost beyond endurance and they narrowly escaped disaster on the island of Olan off the Swedish coast. One observation made by Richardson as they began once more cautiously to re-make sail after a gale, is worth quoting in full. Upon remarking that they overtook and passed a very snuggly rigged Dutchman he remarks, "It is a common saying among Sailors that the Dutch, though they do not hurry themselves in carrying much sail, yet generally get on in their voyage as soon as others." It was a sentiment that was echoed ashore and afloat many times in the succeeding years.

The homeward bound convoy was mustering at Elsinore and under the protecting wing of a naval ship of the line and a sloop, they picked their way cautiously home. But autumn gales forced them to shelter in Norwegian ports

A fleet of colliers off Flamborough Head by Robert Dodd 1797. Fleets such as this, accumulated by foul wind or stress of weather, were a feature of the coal trade. *National Maritime Museum, London*

before a favourable slant drove the *Forester* towards England with her weakening crew. At last anchored in the comparative safety of the Humber they assessed their situation. Their sails were damaged and of the ship's complement of twenty, only eleven remained unaffected by fever or frost-bite and those who could muster for work were weakened by inadequate food, eaten uncooked, for communications with the galley was impossible in heavy weather. Yet the *Forester* was fortunate. Merchantmen who delayed until the last moment, before winter ice and gales closed the Baltic for navigation, frequently kept homeward-bound convoys waiting for the laggards and this resulted in lamentable tragedies. The temptation to extend the trading season as long as possible was a strong one. The expanding ship-building industry demanded a steady flow of masts, timber and flax and those originating from the Baltic were of excellent quality. Importers, often ship builders themselves, were supplying a rising market, the price obtainable for soft-woods had almost trebled during the wars. As part owners of their vessels with interests as shippers, the captains themselves allowed their better judgement to be undermined. In 1808 a convoy was lost trying to force its way through the ice and in 1811 a combination of circumstances saw a tragedy which caused more casualties in the escorting naval vessels alone than were sustained at Trafalgar.

A convoy of 129 merchant vessels, many colliers from the north-east ports, sailed homeward through the Skagerrack delayed by weather and the consequent grounding off shore of their heaviest escort and inevitable late-comers. Less than half of the convoy finally made port and one contingent, driven before a gale down the narrowing funnel of the North Sea found themselves, instead of sounding into the safety of the Thames estuary, as they had assumed when the water shoaled, were in fact cast away upon the sands of the Dutch coast. The *Forester*, however was more fortunate and although her feathers were badly ruffled, could at least discharge her cargo, some of which, not surprisingly was damaged, and then take on ballast at Hull and make sail for Shields. Here the bark was once again rigged down for the winter and the seamen paid off. William, as an apprentice, accompanied his newly-married captain ashore to find employment in his home, assisting with the house-work, an occupation which, rather surprisingly, he does not consider beneath his dignity after such manly adventures. Perhaps the access to a well-filled larder and the attentions of Betsy the maid helped to quell any feelings of indignity.

With the coming of another spring the *Forester* was off to Wyburg, in the Gulf of Finland, as soon as the winter ice was broken. Departing in ballast she returned with deals to Newcastle, the decks piled high with timber above a well filled hold. This was one of the brief periods of peace at sea during the eighteenth century and the most was made of it. Coals were loaded for St Petersburg and safely landed within the mole at Cronstadt and then a cargo of iron, tallow and hemp were

The little port of Cley on the North Norfolk coast imported sea-coals until 1880. The *Tweedside* built 1854, was owned there by Porrit *Roger Finch*

shipped for the *Forester's* home port, the aging bark "having been only three months on this voyage". The succeeding year was employed in a less adventurous fashion, trudging the East Coast with coals for London, but at the end of the season, when the crew was anticipating laying up on the Tyne, an opportunity occurred to load a freight of Stockholm tar at distant Archangel. Although the season was running out fast and the foremast hands deserted, another crew was shipped and the voyage was safely accomplished. While loading at Archangel a Russian gang of caulkers was engaged to caulk the ceiling of the hold and then plaster it over with clay, to guard against the leaking barrels of tar fouling the ship's pumps. This was but one example of the various expediencies that were employed by the masters of small wooden ships in the days of sail to deal with awkward cargoes and leaky hulls. Pine boughs and shrubs were cut as opportunity afforded to line the hold and ameliorate the effect of bilge water spoiling the cargo as the vessel rolled deeply.

There followed a series of voyages with coal cargoes to ports as various as Marseilles, Gibraltar and Cartagena with adventures both aboard and ashore. Food ran short and "oatmeal and flour boiled together like hasty pudding, and some wine poured on top as a substitute for butter" produced some "grumbling at first, but in a few days we got reconciled to it, and liked nothing better". A fellow apprentice falls from the topgallant yard, some seventy feet aloft, but saves himself by clinging to the beef-net towing astern and climbs "up without any hurt except a good ducking". A cargo is secured from a Moorish prince and delivered to the shore by camel and bullock train, gold is smuggled, at dead of night, from Portugal. They meet an Englishman who has shipped aboard a French brig and who admits to eventually appreciating their food, even snails. Finding the *Forester* with only one good pump and the Bay of Biscay to cross, the carpenter fabricates another with nailed battens, tarred canvas and rope. Navigation and the choice of destination have a fine broad approach to them; having loaded salt from the salt-ponds at Cartagena they take their departure "our captain intending to steer for Ireland or America, whichever way the wind answered best".

Eventually William, in 1788, qualified as second mate upon the expiration of his apprenticeship; he now earned the sum of £2. 5s. a month and in this new capacity sailed from Shields, aboard the *Forester,* for Bordeaux with coals. But William's open hearted narrative now shows unmistakable signs that he is seeking wider horizons and a richer variety of companionship and conversation than was offered in the *Forester's* after-cabin on her ocean tramping. He shipped aboard the newly built *Mosely,* serving before the mast and earning £3. all found for the round voyage of about one month. Stoutly constructed the *Mosely* was capable of trading throughout the year . . . . "a fast sailing ship of 15 keels, or upwards, of three hundred tons". Long and narrow the *Mosely* had been consciously so proportioned to defeat the tonnage acts which laid emphasis on beam and upon which dues were paid. She had no bulwarks, only open rails, "so that when laden we could wash our hands over the gunnel" and her decks were hardly ever dry. However, she could be driven harder than the infirm *Forester,* despite these drawbacks, and was worked throughout the year. In 1789 she made thirteen voyages in the London coal trade taking advantage of the high market prices obtainable during the winter months.

After some half dozen more voyages aboard the *Mosely* William was again looking as James Cook did before him, "to see more of the world than dragging about in a collier, where there are no hopes of promotion but for owner's sons and cousins". The sight of a fine armed vessel, copper-bottomed, a rarity then, fitting-out on the Thames, decided him and paying off from the collier he joined the crew of *The Spy,* for such was her name, a slaver. Narrowly escaping the attention of the press-gang, for the wars had begun, he sailed for Africa and passes, as far as we are immediately concerned, out of our view.

The details of day-to-day life and the routine of seamanship aboard the colliers are inevitably ignored by Williamson who naturally dwelt upon the dramatic and unusual. To some extent we can remedy his omissions by studying an inventory of a contemporary collier which has happily survived. It is of the *Gallant,* built in 1788 (Appendix I) and advertised for sale 'carrying near 7 keels of coals, well found and fit for sea immediately'. Fortunately the conscientious notary's clerk made an exhaustive list of her equipment, listing everything from "1 Main Mast" to "3 Mouse traps" and including amongst much else the evidence of some past celebration with a terse "30 Glass bottles".

The *Gallant* was evidently rigged as a snow despite her official designation as a brigantine, a fact we may deduce from the listing of a try-sail mast which, stepped aft of the mainmast, carried the main fore and aft sail. She was well provided and carried studding-sails on the foreyards and "in the hold" were sufficient spare canvas and tackle to meet most emergencies.

But it is the equipment of a domestic nature which throws light upon the life of her crew as it was lived above and below decks. In fair weather at sea the crew were employed tarring the rigging or manufacturing spun yarn, a light line used for temporary jobs aloft. This was produced from old rope with the aid of the

A billander by Charles Brooking 1723-59 showing the fore and aft mainsail which distinguished the type. Many were employed in the coal trade to the smaller ports in the mid-eighteenth century.

From a Private Collection

"spun yarn winch and Bolt". The cabin cooking equipment including a "cheese toaster", a coffee mill, tea-pot and cannister, surprisingly, for tea was a delicacy at this date, two spits and trivet. A certain gentility was established in the cabin aft with the assistance of "3 Cups & Saucers, 1 Mustard pot, pepper Box and Salt Stand" while warmth was supplied from a "Brass Stove". Forward in the fo'c'sle the seamen made shift with a "Fire hearth & Pan, 1 ladle and tormentor, 3 bowls and Platters", but were, no doubt somewhat compensated by recourse to the "3 Iron Bound Beer casks". The *Gallant* was amongst the smaller of the two masted vessels employed in the coal trade, she was about 150 tons burthen, and so, unlike the heavier shipped rigged 'cats' a carpenter would not be numbered in her complement of six or seven. Nevertheless she was furnished with a full outfit of tools including an adze and "6 Gimblets and 1 Pair of Pinchers" besides a "Nut-screw", which we would recognise as a spanner.

The navigational instruments, the "2 Brass Compasses" and "2 Wood Compasses" speak of a well equipped vessel. When in use these were set in the binnacle on deck in pairs to protect them from damage. At this period the binnacle consisted of a substantial wooden box set on legs before the helmsman. To illuminate it at night there was a "Copper candlestick & save all for the binnacle" set between the compasses. Compass deflection and its cause were unknown in the eighteenth century and the effects of carriage-guns, or a return cargo of Swedish iron from the Baltic upon a calculation of the vessel's position must have caused untold bewilderment if not outright disaster. Much of the navigation in the narrow seas consisted of plotting the ship's position by 'dead-reckoning' and to calculate this a knowledge of the speed was essential. The "Logline and Reel", and "2 Half Minute Glasses" and "2 Quarter Minute Glasses" and "1 two hour Glass" indicated how this was done, for there were no chronometers aboard small merchantmen in the closing decades of the eighteenth century. The speed was calculated each half-hour of the watch by casting a chip of wood overside at the bow and timing its progress along the vessel's length with the assistance of the quarter-minute sand-glass.

To ensure greater accuracy it was usual to cast the log and time the run of its line, knotted at intervals to mark its length, with the small time-glasses, from which the vessel's speed was calculated. The nautical measurement of speed in knots derives from these marks on the line. The recording of this information was facilitated with the assistance of a traverse board which had been in use amongst the seamen of north east Europe from the early sixteenth century. It had the advantage of linking, in one simple recording instrument, a record of the estimated distance sailed and the related compass bearing. Eight pegs were placed consecutively in holes drilled in the compass-rose scribed on the board; each peg indicating the bearing of the course sailed each half-hour. At the base of the circular portion of the board the speed was pegged into a scribed grid, at half-hour intervals

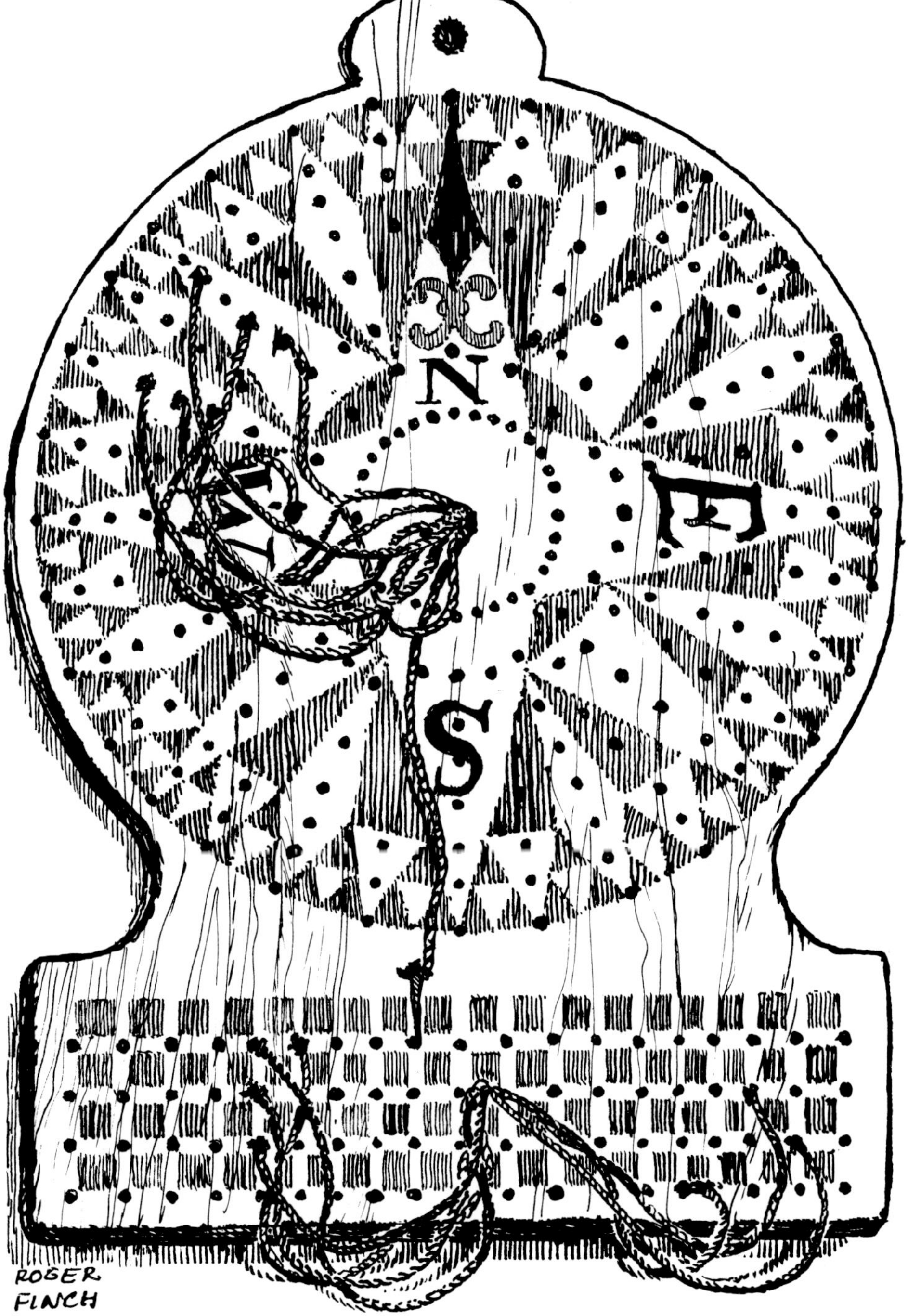

Wooden traverse board, pegs were of lead or bone. *Drawing Roger Finch. Hull Maritime Museum*

throughout the watch. Even a largely illiterate crew could in this way record the information necessary for the captain to prick off the ship's track on a bearing and distance chart at the end of the watch. A bell, perhaps the 'Cabin Bell' listed, which was struck at each turning of the sand-glass during the watch, told the crew of the passing of time.

The lead and line, of which the little *Gallant* had two, one for sounding in-shore and one for deep-sea work, was another essential and perhaps the most ancient instrument of navigation. Its correct use (and to do this effectively the leads-man stood in the chains, lashed to the shrouds for safety), was the hall-mark of a first class seaman. It was with the lead-line that, by tradition and in fact, the collier-fleet smelt its way up and down the North Sea. Not only did it gauge the depth of water beneath the keel of the collier, it also furnished a method for discovering the nature of the sea-bottom for the hollowed out base of the lead was 'armoured' with tallow to which material from the sea-bed adhered, for inspection and identification. An early chart of the North Sea, dated 1777 and appearing in a somewhat optimistically entitled publication, "A Complete Channel Pilot", marks amongst many others in appropriate positions, "These soundings are mixed with Fine Shells", "Light brown soundings" and off the Galloper, in the entrance of the Thames Estuary, "Course soundings mixed with Scollop Shells and Stones". When crossing the North Sea it was said that if the pebbles brought to the surface could be broken with the teeth one was westward of the Dogger, and if not one was to the eastward.

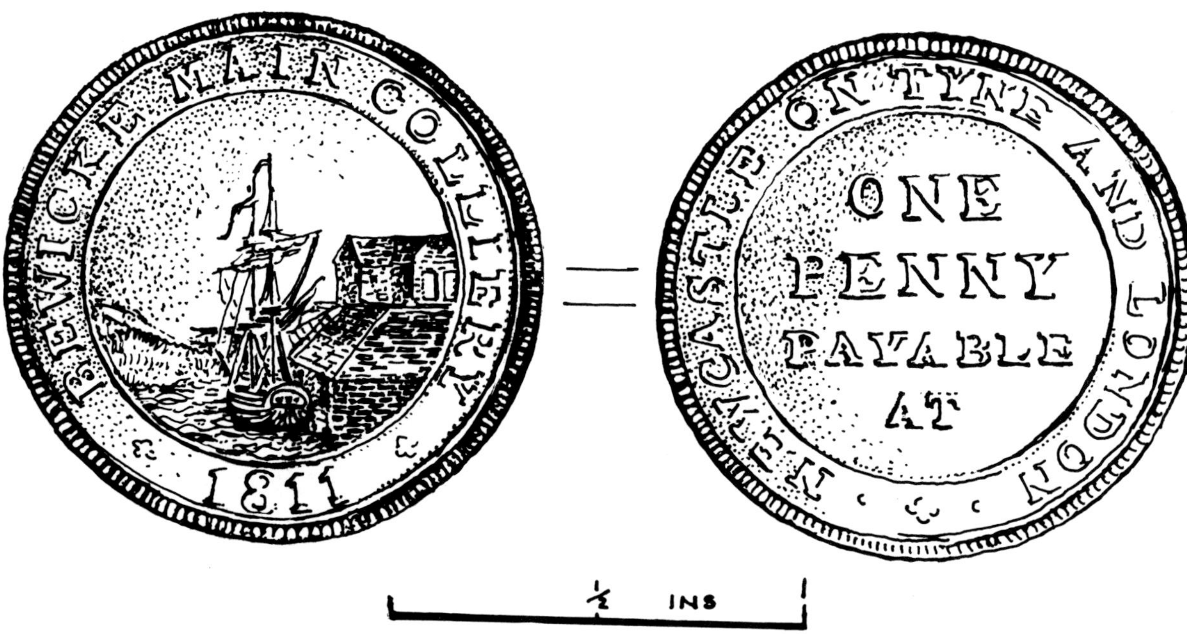

8 Wider Horizons

THE great majority of the vessels which made up the collier fleets were launched, traded and finally left their bleached bones on a ship-breaker's yard or a storm-wracked beach without any record of their existence except terse entries in faded port-books or official lists. Of the few escaping this anonymity, those employed in the eighteenth century to explore and chart the South Pacific are particularly memorable. Like their crews who were drawn from the ranks of the Navy, they had greatness thrust upon them and rose to the occasion well. While we know little other than what we can deduce and infer about the earlier ships of discovery, those of Cook and his followers are fully recorded for they were merchant vessels bought into the Navy. By the eighteenth century vessels acquired by the Admiralty and modified under their supervision or upon their instructions, were measured and recorded by official draughtsmen with a commendable thoroughness. Indeed, there is evidence that the care given to drawing up the plans was not always matched by the watchfulness and honesty with which the necessary alterations were carried out. However, the system, whatever might have been its shortcomings in practice, and these were an endemic failing of the times, enable us to reproduce here the hull plan and lines of a typical collier in the heyday of the coal-fleets. For it was as the *Earl of Pembroke,* a "cat-built bark three years ten months employed in the coal trade" that the vessel which was to become James Cook's famous *Endeavour* entered the Navy in 1768.

Three other colliers had been inspected before the Whitby-built *Earl of Pembroke* was selected and purchased for £2,800. As there was already a sloop *Endeavour* of fourteen guns in the Navy Lists at this time the re-named collier was officially entitled *Endeavour Bark.* Cook himself had requested that his proposed expedition

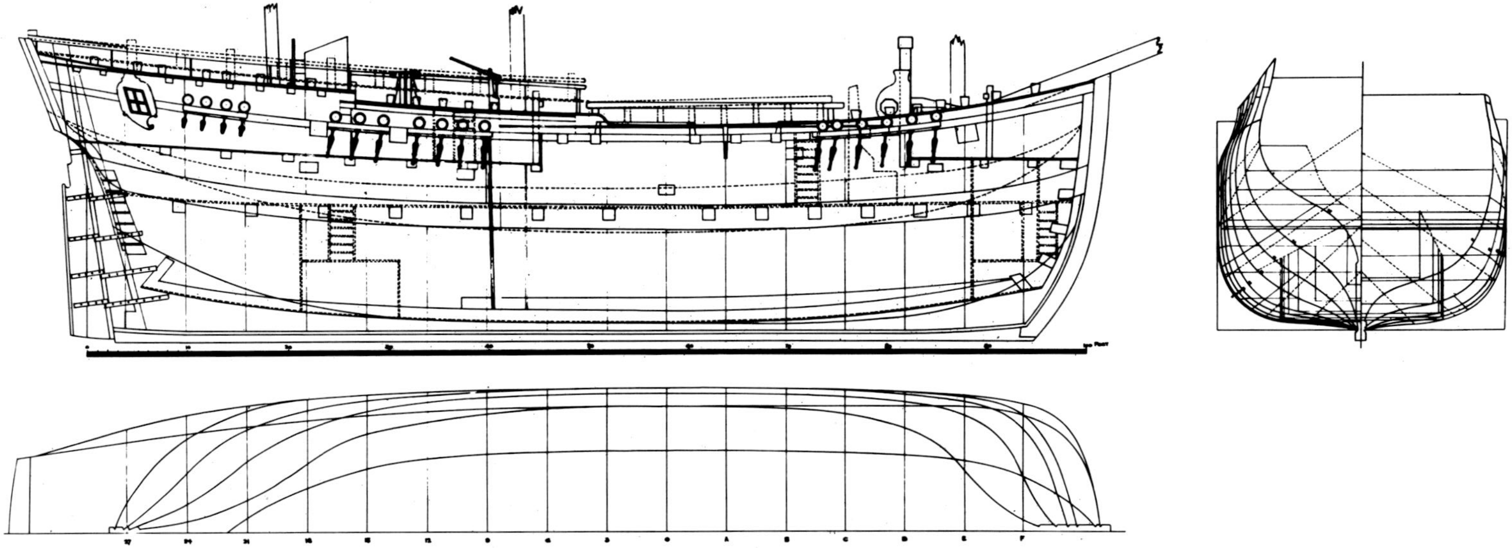

Draught of H.M. Bark *Endeavour* 1768. She had been built as the *Earl of Pembroke* at Whitby in 1764. It accurately records the lines of a collier bark. Her body plan taken off the single deck. Length of the range of the Lower Deck—97 ft 6 in.; Length of Keel for Tonnage—81 ft.; Breadth extreme—29 ft. 2 in.; Depth in hold—11 ft. 4 in.; Burthen in tons—566 49/94. *Science Museum, London*

should be undertaken with a North Country collier. His choice was dictated by the knowledge that to mount successfully an expedition into unknown seas remote from assistance, survival depended first and foremost upon the vessel's ability to survive and an unpretentious collier bark had qualities which would stand the test better than any other. There were bigger and faster vessels which, with the official backing he enjoyed in influential circles, Cook could have commanded. But sheer size, while it could accommodate the stores, scientists, and equipment which accumulated in overwhelming abundance, brought its own problems. Large vessels had a deep draught which would make navigation in shallow waters hazardous; the heavy spars and complex rigging would be difficult to replace and hard to work, for the crews were inevitably decimated by disease as the voyage progressed. A vessel of the tonnage of the collier could in an emergency, and voyaging the oceans on an expedition lasting years was one series of emergencies, prove more easy to handle. It could be manoeuvred by its own boats towing ahead, assisted by the crew manning sweeps worked through ports in the hull. Spars and even masts could be cut and trimmed in forests from native timber and then stepped in place aboard without recourse to elaborate dockyard equipment.

While a frigate would have been faster, sail marginally closer to the wind and make less leeway, the relatively sharp rise of her floors amidships meant she would be more vulnerable to damage should she go aground. To colliers this was all part of their workaday life and they were constructed with the certain knowledge that they would receive the hardest treatment as a matter of course.

For a fuller account of James Cook and his achievements I must recommend the reader to one of his numerous biographers, but perhaps most strongly to his own journals kept at sea and covering the years of adventure 1768-79. In passing, one should say that perhaps the fascination of his personality lies in the fact that a man who combined so many talents with such determination, courage, and native commonsense, remains finally an enigma. He was born at Marton, in Cleveland, Yorkshire, and on '1728 Nobr. 3, James ye Son of James Cook day labourer baptized" was entered in the register of Marton Church. Cook was undoubtedly fortunate that, despite his humble origin, he secured a basic knowledge of reading and computation. At the age of sixteen he became an assistant to the haberdasher and general dealer at Staithes, a fishing village set in the cliffs near Whitby. Cook, however, found that the vending of gingham and gingerbread was not, after all, to his liking. He appealed to his employer to assist him to find an alternative and he, good man, rode with the youth to nearby Whitby where he helped him to secure an apprenticeship with Mr John Walker, a Quaker and shipowner of the port.

Walker, like most north-country shipowners of the time, was also a master-mariner, and perhaps he recognised in the eighteen-year old grocer's assistant some of the latent qualities which were to be of equal value, whether employed sounding a loaded collier through the Barrow Deeps or later, and in more distant waters, plotting the Australian Great Barrier Reef.

His first ship was the *Freelove,* John Jefferson master, and Cook was entered upon the muster roll as "servant". The *Freelove* was 450 tons and she must have been one of the largest in the coal-trade at that time. After making several coal voyages in this lowly position, for 'servant' indicated that he was neither an indentured apprentice nor a seaman of enough experience to take his trick at the ship's wheel, he transferred to the *Three Brothers.* The *Three Brothers* was a new vessel and Cook assisted in fitting her out although as a "servant" it must have been a wearisome business of tarring and scraping, stowing ballast, and filling harness casks with beef and brine. Then, when in all respects ready for sea, she loaded coal and sailed for London. However, upon arrival, under the command of her master-owner, the *Three Brothers* was hired to the Government as a cavalry transport and fitted with stalls for 40 horses, built upon a deck laid over the ballast. The colliers, like their crews, were considered a valuable reserve of ships and men to be drawn upon in case of war or rumours of war. Unlike their crews, the ships were private property and their requisitioning was well paid for and adequately recompensed in the event of loss; the men from the colliers when pressed for the Navy were far less considerately dealt with.

In 1749 Cook was rated as a seaman and in the same year the *Three Brothers* reverted to private trading. Ship-building was booming at Whitby and there was a growing demand for Baltic spruce and fir so that between May and September colliers would trade to Norway for timber. The *Three Brothers* participated in this, alternating voyages across the North Sea with coasting before the winter closed in.

It was still the custom in the second half of the eighteenth century to lay-up colliers for the stormy, dark months of January and February when the bar at the entrance of the Tyne became impassable for days at a time. During this time, while they were re-fitted and the rigging was overhauled, it was usual for young seamen and apprentices to lodge ashore with their masters. Working during the day, splicing, serving ropes, and sending up masts and spars, the long winter evenings gave Cook the time for study, and he made the most of the opportunity. He worked hard to expand his knowledge of arithmetic, which was sufficient for an aspiring grocer, to encompass the mathematics required for navigation and surveying. In 1752 his promotion to mate, after further experience in a Stockton vessel, at the age of 24, of the *Friendship,* also owned at Whitby, still found him in the North Sea and coasting trade. The almost uncanny reputation which Cook acquired at this time for a capacity to choose instinctively the correct heading and to be aware of impending danger before anyone else, is not necessarily a extraordinary one. It was achieved by a mind which was ever absorbing and recording, awake or asleep, all the subtle changes of wind forces, of distance run, of lee-way made, of the tidal streams, of different helmsmen's idiosyncrasies and individual vessel's sailing characteristics. Computing these with a dispassionate mind made for a reputation which was unequalled, but to a lesser degree was a skill acquired by all master mariners in the days of sail.

While his ship was lying in the Thames in 1755 Cook made the decision which was to bring him to the quarter deck of the *Endeavour* and finally to a tragic death in Hawaii. He requested his discharge from the *Friendship* and joined the Royal Navy as a volunteer, "having in mind to try my fortune that way" as he laconically expressed it. Perhaps the tight-lipped convergent in Cook's personality saw in the discipline, hierarchy, and order in the Navy something which instinctively appealed to him; most collier-masters were share-owners in the ships they commanded and Cook had little chance to acquire even modest capital, and had few expectations from his parents. He had no Quaker bankers amongst his relations, as had many aspiring shipping families, although, it is said the Walker family had promised him a command before his last voyage to London.

Serving in North American waters Cook found a widening scope for his developing talents as navigator and surveyor. So successful was he in this work that he won the rank of Master, the title given to an officer responsible for navigation. He had the reputation of a skilful surveyor, unspectacular but indefatigable in his industry. Carrying a fulsome recommendation to the Admiralty from his commanding officer, a man who himself had entered the service as a volunteer from the Merchant Navy, he returned to England in 1767. His return home coincided with the need for a new breed of adventurer – the seaman-scientist – and in April, with the support of the Royal Society, Cook was appointed to command an expedition to the South Pacific, ostensibly to carry a party of scientists to undertake astronomical observation, but in actual fact as his secret instructions commanded, intended to explore and seek for continents and islands and "with the Consent of the Natives to take possession of Convenient Situations in the Country in the Name of the King of Great Britain".

The choice of the bark *Endeavour* by Cook for his voyage we have already discussed. She had been built by Thomas Fishburn for Captain Thomas Milner of Whitby and while no contemporary sail plan of the vessel survives we have a painting, now in Australia, by Thomas Luny, of the ship leaving Whitby as the collier *Earl of Pembroke.* Luny was only a child at that date so even this picture must be something of a pictorial reconstruction of the event. We may, however, with confidence consider the colliers as almost identical, in both hull and rig with the *Endeavour.* Any rigged model of this ex-collier derives its dimensions and proportions of the spars, masts and rigging from contemporary manuals printed for seamen and shipwrights, supported by evidence from prints and paintings, for no sail plan survives from either her naval or later career. Such modifications to the collier as were required for the expedition were carried out at Deptford and she sailed from the Thames some four months later. The time in the dockyard was largely spent in contriving accommodation for the greatly increased crew enlisted for the expedition. As a collier her complement, ranging from captain to newly recruited ship's boy numbered about 14, certainly less than a score. Now some 94 persons, not to mention Mr Joseph Banks, the dilettante scientist, his pet greyhound and a goat which had already circumnavigated the globe aboard the *Dolphin* of Captain Wallis, had to be accommodated. Besides the building of cabins and decks, other equally essential work had to be carried out on the hull of the *Endeavour* to guard against attacks of the *teredo navalis,* the boring worm particularly active in tropical waters. To provide some defence against this threat the bottom of the vessel was 'swathed and filled' with extra planking into which were driven hundreds of large, round-headed nails.

The voyage really began from Plymouth in June 1767 where the *Endeavour* collected the scientists who accompanied the expedition. Opportunity was also taken to construct a platform on the poop to enclose the steering gear and give a convenient position from which to take sightings when surveying was in progress. Despite Cook's quietly efficient leadership the long voyage was not without incident. At the outset the ex-collier's austere appearance and particularly the fact that she retained the plain upright stem of her kind, without figurehead or head-rails, led to the Viceroy of Rio de Janeiro refusing to believe in the expedition's official and declared purpose. Only considerable forebearance on the part of Cook enabled them to leave with the replenished stores they urgently required. The rounding of Cape Horn was fortunately uneventful but the nearest the *Endeavour* came to total loss was on the Great Barrier Reef, off the Australian Queensland Coast.

In bright moonlight the little ship struck and ground her hull on the hidden, lethal-sharp coral reef. By the jettisoning of water, ballast and finally guns; by using every subterfuge of seamanship and the physical strength of the entire ship's company to its limit, the *Endeavour* was shifted off and limped clear into deeper water. The leak in the wooden hull left by the encounter could scarcely be controlled by pumping, but was dealt with by 'fothering'; an expedient one cannot resist explaining in Cook's words, from his own journal.

> . . . . . "some hands employed sowing ockam wool etc. into a lower studding sail to fother ship . . . . . we got the sail ready fill'd for fothering. The manner it is done is thus, we mix ockam and wool together (but ockum alone would do) and chop it so small and then stick it loosly by handfulls all over the sail and shrow sheeps dung and other filth. Horse dung for this purpose is best. The sail thus prepared is hauld under the Ships bottom by

> ropes while the sail is under the Ship the Ockum etc. is washed off and part of it carried along with the water into the leak and in part stops up the hole. Mr. Munkhouse one of my Midshipmen was once in a Merchant ship which sprung a leak but by this means was brought home from Virginia to London to him I gave deriction of this who exicuted it very much to my satisfaction."

Alas, young Mr Munkhouse never reached England to tell of his commander's faith in him and his vindication of it. He died from fever, contracted at Batavia, on the long voyage home. Unlike Mr Banks and his fellow botanist Solander, who were similarly afflicted with the infection, he was not in the position to purchase a Malay Woman and resort to a country house "hoping that the tenderness of the sex would prevail here, which indeed we found it to do". Munkhouse was singularly unfortunate, for the humanity and scientific understanding which Cook brought to the problem of maintaining the health and morale of those under his command was greatly ahead of his time, for disease and malnutrition at sea was tolerated with fatalistic acceptance in the eighteenth century.

It seems scarcely possible, but so accurately did Cook plot his position when he rubbed shoulders with disaster on the Barrier Reef, that in 1969 an expedition from the Academy of Natural Sciences, Philadelphia, U.S.A. recovered the cannon which had been jettisoned two centuries earlier. Restored to its original appearance by the Australian Government, one was presented to the people of Great Britain and is now in the National Maritime Museum, Greenwich; a salvage operation and outcome which, one feels, would have met with Cook's entire approval.

The *Endeavour* dropped her anchor in the Downs on 12th June, 1771, in home waters again almost exactly three years after leaving Deptford. Cook received the honour which was his due and the ex-collier was refitted for further service in the Navy. She was dispatched to the Falkland Isles as a storeship and then, after being sold out of the service in 1775 the *Endeavour* reverted to her role of an East Coast collier. In 1790 the now ageing bark was purchased from a firm in Dunkirk and under French colours as *La Liberte.* In 1793 she sailed into Newport, Rhode Island, with a cargo of whale oil and bone, ran aground when leaving and upon inspection proved to be rotten and was condemned; however fragments of the hull were built into new ships while some were retrieved by later generations of Rhode Islanders and proudly treasured.

The *Endeavour* was unavailable for Cook's second voyage of exploration to the Pacific in 1772. But he was still faithful to his choice of Whitby-built colliers and secured the use of the *Marquis of Granby,* 462 tons, which was renamed *Resolution* and the smaller *Marquis of Rockingham* whose name was changed on purchase to

The *Endeavour* of Captain James Cook. A model in Whitby Museum, showing stunsails set on the fore and main mast.

Adventure. The bigger vessel cost the Admiralty £4,151 whilst the 336 ton *Adventurer* was bought for £2,103. Both were converted for naval use at the Deptford yard. The *Resolution* was fitted-out with a frigate's headrails and figure-head as an embellishment to her plain collier-cat's stem. It was felt that a vessel in the service of his Brittanic Majesty might not receive the due degree of respect without this ornamentation to her bluff bows; Cook himself had been caused some embarrassment at Rio when, on his first voyage, the port authorities had doubted the official status of the *Endeavour* with her simple unadorned North Country bows.

Other modifications of the *Resolution* were directed by Joseph Banks who was determined not to undergo a repetition of the cramped conditions he experienced aboard the *Endeavour.* The results were almost catastrophic. The vessel was transformed from a seaworthy, if slow, ocean voyager to a cranky hulk, hardly capable of venturing beyond the Nore. Cook remedied the situation by stripping off the top-hamper that had been added and the *Resolution,* a considerably larger ship than the *Endeavour,* served the expedition and its accompanying scientists well in the North and South Pacific.

For Cook's third and last voyage of 1776 he commanded the *Resolution* but the *Adventure* proved to be unavailable and was replaced by yet another Whitby collier, the *Discovery,* of a little under 300 tons. The *Discovery* set out later than the *Resolution,* now a sadly worn vessel, hastily and inadequately refitted, and the two vessels met up at the Cape where the explorer had been doing his best to remedy his vessel's deficiencies. They sailed through the Pacific and its tropical islands to the Arctic ice and upon their return James Cook met his tragic end. Murdered at Hawaii, worn out, like his ship, by the sea and faithful service; "the ablest and most renowned navigator this or any country hath produced". In October 1780 the *Resolution* arrived in the Channel after a voyage of over four years. Re-fitted she served on in the navy as an armed transport and was sent to the West Indies while the little *Adventure* of the second voyage returned to trade and owners at Whitby. She lasted until 1811 when she came to grief upon the rocks of the St Lawrence River.

Yet another collier bark, the *Xenophon*, earned renown as a ship of discovery far beyond the grey North Sea but is less well known than the *Endeavour.* Re-named the *Investigator,* the *Xenophon* under the command of Matthew Flinders, sailed in the wake of Cook to the Antipodes and became perhaps the most famous ship, after the *Endeavour,* involved in the discovery and charting of the Australian coast. Like her celebrated precursor, the plain workaday collier bark *Xenophon* was purchased by the Government, but unlike the *Endeavour* she was not originally intended to carry an expedition of discovery to distant seas. The *Xenophon* was bought for conversion to an armed sloop and to act as a patrol or convoy-escort vessel. The opening years of the Napoleonic wars with their insatiable demand for this humble but essential form of armed vessel, made ever more pressing as French privateers multiplied, had forced the Navy to accept this expediency. The encouragement of a large collier fleet, to develop as a reserve upon whom the navy could draw, was accepted government policy, but once 'bought in' the barks usually served as transports or auxiliaries rather than as gunned vessels. While her career turned out to be an unusually eventful one, the successive stages in the story could be parallelled by many other colliers "taken up by the navy".

The *Xenophon* had been built at Monkswearmouth Shore, Co. Durham in 1795 by Henry Rudd. He must have had a taste for classical names, for amongst others he launched were the *Argo, Cicero* and *Omega.* Many ship-builders, not primarily ship-owners, built vessels as a speculation and then traded them for a short period until they found a purchaser and *Xenophon* began her career, owned by Rudd, in this way. She made a few voyages in the coal-trade under her North Country owner and then passed to a coal-fitter of Co. Durham; a contemporary entry in the register describes her as a three-masted, square-sterned ship with no head or gallery and of 328 tons. She was purchased by the Government after three years, in 1798, while lying in the Thames for £2,900, although the surveyor had valued her hull at £2,414 and her masts and spars separately at £116. This second figure excluded sails and much equipment which, for a collier in trade, would have probably trebled the amount. However, if their Lordships could have foreseen the length and extent of the *Xenophon's* future career they would not, I feel, have been displeased with the bargain they struck. But a collier, however sturdy her construction and well suited to her trade, required extensive modification before she would be capable of taking up her new duties for the Navy.

This reconstruction was undertaken at a private dockyard at Deptford and while the exact extent of the work is problematical it is certain that when she emerged as His Majesty's sloop *Xenophon* she carried eighteen 32 pounder carronades and two 18 pounder guns "for chace". Carronades were light guns, limited in their range; nevertheless they were accurate and particularly suitable for smaller naval vessels and merchant men and were in use aboard the colliers. Their relatively light weight did not place too great a strain upon the fabric of the vessel.

Now rated as a sloop-of-war we can also be certain that *Xenophon's* traditionally austere appearance of a North Country collier had been embellished forward with a frigate's projecting stern, head rails and figure-head, while aft, quarter galleries had been constructed to illuminate the accommodation under the quarter deck and provide lavatory arrangements for the officers. A deck had been laid in what was once the hold to carry the main armament and furnish living space for her crew who slung their hammocks between the carronades. Above this, the quarter-deck was now protected by substantial bulwarks and over the stern, above the stern gallery, projected the two 'stern-chacers'.

In 1799 she was stationed at the Nore and now presented a formidable deterrent to any French privateer who might be seeking victims amongst the fleets of merchant-men who converged on the Thames estuary.

Such was the ship, what of the man? At this time Matthew Flinders was seeking to secure support for an expedition to fill a blank on the outline map of Australia, a blank left by previous explorers of the sub-continent. Although a young naval officer of only twenty-six he was a veteran of the Southern Ocean and had served in an expedition led by the redoubtable Captain Bligh, who had served under Captain Cook. A direct link between the career of Captain Cook and the young Flinders existed in the person of Sir Joseph Banks, who had accompanied the Whitby born explorer aboard the *Endeavour.* Eager to re-visit the Pacific, Banks gladly supported the young Flinders' enthusiasm to complete the work of his illustrious predecessor along the Australian coast. Flinders used the scientist's extensive influence to further the project of a voyage of investigation and discovery in the uncharted seas. As a result of Banks' pressure the Navy agreed to provide an appropriate ship for the purpose of exploration and the *Xenophon* was chosen and renamed *Investigator.* Fortunately Banks' interest in the practicalities of the voyage ceased at this point. He did not influence the re-fitting of the sloop, as he had the alteration of the *Resolution* of Captain Cook some twenty-five years earlier, with such disastrous results.

To prepare for the long voyage south further modifications to those already carried out were found to be necessary to the hull of the one-time collier. Cabin space had to be provided for the "five men of science" who were amongst the members of the expedition and also storage room for the specimens that it was hoped to collect. Much else of a maritime nature had to be provided; davits on the quarters for the extra boats carried for survey work, hammock stanchions amidships to stow the rolled bedding of the crew; four gun-ports were replaced by windows to light the cabins and the four now redundant guns were sent ashore. The somewhat improvised accommodation was by no means lavish for a group of men who were to work and live together for months on end. But compared with the conditions aboard the *Endeavour* and later expeditions such as that recorded by Darwin on the *Beagle,* it was extensive enough to reduce the likelihood of friction to a minimum. Nevertheless, by the time *Investigator's* ballast had been replaced by 35½ chaldrons of coal she must have been an overloaded ark. Such was the shortage of space that when the ship finally sailed from Spithead on 18th July 1801 the spare rudder and trawl-heads, for gathering specimens of marine life, had to be lashed out-board in the chains, where the main rigging was secured to the hull.

Where the *Endeavour* had been protected below the water line by large headed nails, copper sheathing was now more common and this had been carried out when the *Xenophon* had been reconstructed at Deptford. The lesson had been brought home forcibly by the examination of the unsheathed lower hull of the *Endeavour,* at Batavia, upon her return voyage. It was discovered that "the worms had made their way quite into the Timbers, so that it was a Matter of Surprise to every one who saw her bottom how we had kept her above water".

To learn of Matthew Flinders' long voyage and his scientific investigation into the cause and effect of compass deviation the reader must look elsewhere. Suffice to say here that they combined fine professionalism, high adventure, acute and painstaking scientific observation with a humanity which only occasionally relieves the story of nineteenth century imperialism. H.M. Sloop *Investigator* unfortunately scarcely equalled the calibre of her young commander. She leaked. All wooden vessels leak, but the *Investigator* leaked with a persistence and malignancy which spoke to her carpenter and master of something far more severe than a minor, easily traced defect. The resources of skill and materials and the ingenuity of the picked crew that manned naval vessels in the eighteenth century made possible miracles of improvisation; but there was a limit even to their resources. Beyond a certain point a dockyard and its specialised equipment were essential. As a result of the deterioration to her hull after circumnavigating Australia in 1803, the *Investigator* was condemned as unseaworthy and became a storeship at Port Jackson, where the first free settlers were establishing a colony. The shortage of shipping resulting from a long war made it increasingly difficult for colonists to maintain links with neighbouring settlements. Permission was sought from the Admiralty to re-habilitate the hulk and use her to augment the colony's tenuous links of communications. The required permission was granted and, as so often happens, it was discovered that the original builders of the *Investigator* had worked honestly and well and the integrity of Mr Rudd and his North Country shipwrights was vindicated. The work of alteration undertaken at Deptford was of a different calibre with the result that rot had infected the reconstructed topsides, endangering the whole fabric.

With commendable thoroughness the upperworks were torn off and re-decking and re-caulking made a water-tight, but decidedly cut-down vessel of the *Investigator.* Her main deck was now scarcely five feet from the water while her three masts were reduced to two. Enough of the original spars and rigging remained for at least a caricature of her former rig to be set up. Fitted out as a stunted brig, with no canvas above her topsails left to set or even spars available to bend them to, she cruised in Australian waters and then, carrying important dispatches, returned to England. She completed the voyage arriving home in 1805, to be condemned by the Navy Board and to be sold out of the service. But such was the demand for tonnage, with war at sea making continual inroads upon shipping, that she found an eager buyer and moreover one who was prepared to pay a figure not much less than half her original purchase price.

How well did those North Country shipwrights build! Once more named *Xenophon,* she began a third career, now owned at Maldon, Essex. She had dockyard repairs which included fitting iron stanchions in the hold; at last the shortage of suitable traditional materials had broken down the conservative attitude of shipwrights and they were prepared to use iron reinforcements to wooden hulls. She was now registered as a brigantine, with a quarter-deck. The early nineteenth century registers were only just beginning to regularise the description of specific rigs and when she was recorded in 1812 as a brigantine this probably indicates she was two-masted with some square canvas on each. Later the *Xenophon* returned to her native North East Coast and was employed deep sea in the timber trade from North America. Perhaps it is just as well, when one considers her age and adventures that she was carrying buoyant cargoes. The old collier was now owned at Newcastle and later by E. Young of Shields who continued to send her to Quebec and to the Baltic. The veteran was now sheathed and patched, the primitive hand-spike windlass of her earlier collier days had been replaced by a patent, pump handled one, but she was still copper-sheathed and this enabled her to carry coal to the warm waters of the Mediterranean without danger of worm attaching the underwater hull. Her rig was now recorded in official lists as that of a snow and this remained unchanged until the end of her trading days. Finally in the eighteen fifties as an old collier she sailed out once more to Australia and was re-registered at Melbourne in 1860. Once again reduced to a hulk at Port Philip, she was broken up in 1872. This was the Port Philip which Matthew Flinders, aboard H.M. Sloop *Investigator* had been instrumental in discovering so many years before.

9 Wrecks and Rescue

"I was surprised to see, in all the way from Winterton, that the farmers and country people has scarce a barn or a shed, or a stable, nay the pales of their yards and gardens, not a hogstye, not a necessary house, but what was built of old planks, beams, wales and timbers, etc., the wrecks of ships and ruin of mariners' and merchants' fortunes"

Daniel Defoe, *Journey,* 1724.

IN 1692 a fleet of colliers, 200 strong, left Yarmouth Roads with a fair wind and were suddenly assailed by a violent gale from the N.E. After passing Winterton Ness, off the Norfolk coast, some of them tacked and managed to arrive back to anchor once again in the lee of the Scroby Sands. The remainder of the company pushed out to sea but did not weather the Ness to the southward. In an excessively dark night 140 sail were driven ashore and scarcely any of the crews were rescued. At the same time a number of coasters from Lynn and Wells, bound for Holland with grain weighed and got under sail from the Roads. They too met with disaster so that some 200 ships and some 1,000 lives were lost in one dreadful night.

Previously in September 1671 a great storm destroyed 75 vessels, most of them colliers, lost on the East Coast. There were to be tragedies as harrowing as this and a great many of them, before an inter-linking system of protection, warning, aid and rescue was to be evolved. The coal-trade traversed a coast which had at least the advantage of a prevailing wind which blew offshore. Moreover the shallow sands, dangerous in themselves, gave good holding ground for anchoring providing their cables held. But largely unmarked until the nineteenth century the shoals were a hidden hazard that could become lethal in thick weather, particularly when the tide set athwart them. By tradition a fleet of colliers, moving together through a narrow channel such as Yarmouth Road, or the Swin Spitway would have a man stationed in the chains. The leadsman would sing out the soundings loud enough for his companions to hear so that craft would be aware of those vessels approaching shallow water; those announcing deeper water would be safely in the channel. The Newcastle fleet would prefer to turn to windward, which with the prevailing South Westerlies they frequently did, rather than run, heavily loaded through the shallow channels of the Estuary. As the soundings shallowed they could go about, a far easier manoeuvre than to round-up in the rush of a head-long, down-wind run. But the tide and the anchor were as important to the collier as wind and canvas. The idea of turning to windward over a foul tide was only attempted when in ballast or when anchoring was impossible, as it was north of the Wash.

Crossing the treacherous Tyne bar brought disaster to many colliers and this hazard was only partially mitigated when the breakwaters were built in 1859, below the ruins of Tynemouth Priory and on the south from the Herd Sands, as part of general improvements to the river. It was not unknown for brigs to be brought in by pilots on their beams end, carrying such a press of canvas that the vessel was forced over and hence drew a vitally few less feet of water. If to stay outside meant almost certain disaster, to find shelter in this manner was a lesser evil. The problem of ballast disposal on the Tyne (see Chapter 10) tempted captains into instructing their crews to dispose of as much as was safely possible, under favourable conditions, while still at sea, but within sight of their destination. Inevitably a sudden unforeseen deterioration in the weather would drive the unmanageable vessel, with insufficient ballast, on to a lee shore; barometers were not found aboard sailing coasters until the end of the nineteenth century. Tragedies did occur through human failing often enough, as when the wreck of the *Freelove* in 1746 was recorded in a ledger tersely as "Lost on 31st December Wm. Brough, master, upon the Hird Sand. All hands lost but the cook who was saved upon a plank; drove from her anchor when she was riding near the Black Middings, for want of a good look out all these lives". But then judgements are easily made at counting-house desks. Another drama, this time with a happier outcome, is hidden in a brief ledger entry recording the loss of the collier bark *Richard,* October 1763 "obliged in a gale of wind to run her on shore at Robin Hood's Bay – all hands saved". It is hoped that the crew then received the sympathetic treatment they deserved; the fact that this sympathy was not always forthcoming is underlined by an Act of Parliament of 1752 stipulating the death penalty for "Plundering any vessel . . . . Preventing the escape of any person that endeavours to save his life; or wounding him with intent to destroy him". But the beginning of a new attitude to events that had been considered inevitable was at hand. A few years later an Act provided the basis of a more positive approach. "In order to assist the distressed and prevent the scandalous illegal practices on some of our seacoasts it is enacted that all head officers and others of honour near the sea shall, upon application made to them summon as many hands as are necessary and send them to the relief of any ship in distress on forfeiture of £100".

The Act had to lie unregarded for many years before it was to be given practical expression. When this finally occurred it was on the North East Coast where the plight of colliers fighting for their lives and losing an unequal battle was clear for all to see. At least the seamen had their dangers and sacrifice well recorded in the contemporary newsheets. Their brothers in the mines were less fortunate. Not until the great disaster at the Felling colliery in 1812 was a gentleman's agreement between owners, magistrates and the press effectively to silence news of accidents and their resulting casualties broken by the Rev. John Hodgson. It was another local Northumbrian clergyman, Archdeacon Sharp, who, in 1786, ordered from the inventor Lionel Lukin, of Dunmow, Essex an adaption of a local coble into an "unimergable boat"; an adaption principally contrived by lashing casks under the thwarts. It was the first life-boat. Close by at the "station of our coal fleets", South Shields, William Wouldhave and Henry Greathead, a boat-builder, jointly designed the *Original:* incorporating the ideas of the latter it was built at

W. Elmes painting and engraving of the *Original* in a rescue incident off Tynemouth Castle. It was published in 1803 by Henry Greathead, builder of the first lifeboat. *R.N.L.I.*

his boat-yard in 1789. The *Original* was instrumental in saving hundreds of lives from colliers ashore on the Herd Sands and the rocks nearby between 1790 and 1830. Whitby had a similar boat in 1799 and another early Greathead design was the Redcar lifeboat *Zetland,* built in 1800. It was finally withdrawn from service after eighty years of life-saving. These pioneers were essentially pulling boats, very beamy, self-righting, and with an enormous sheer. A substantial belt of cork along the gunwale gave added flotation and they were steered by the coxswain with a single oar. Their range was limited to the physical resources of the crew in what were inevitably extreme conditions. Nevertheless they multiplied all along the North East Coast, each coastal community forming its committee, ordering a boat and recruiting a crew from amongst the local longshore men or watermen. The provision of the *Original* of Greathead's building had been inspired by Sir Cuthbert Heron, of South Shields, witnessing the loss of the collier *Adventure* on the Herd Sands, during a gale in 1789. Within only 300 yards of the shore and the sight of a large crowd the entire crew had perished. Local boatmen, protesting that the heavy seas made rescue impossible had refused to launch their own craft despite offers of rewards. As a result of the tragedy a committee of shipowners of South Shields opened a subscription and commissioned Henry Greathead to produce a boat that would command the confidence of local longshoremen and be capable of making rescues from stranded vessels. By 1802 it was claimed that so successful were Greathead's lifeboats that at least 300 persons had been brought ashore from vessels in distress off Shields alone. The Duke of Northumberland had presented a similar boat to North Shields in 1798 and Greathead lifeboats were stationed overseas at Oporto, presented by the Duke, and Memel. Unlike many who benefit humanity to such an extent, Greathead reaped recognition and reward in his own lifetime and in 1802 he was granted a gift of £1,200 by Parliament. Not until 1825 was a National Life Boat Institution formed which could rally inland support and it was many years before all the local committees brought their boats under its auspices.

The dangerous approaches of the Thames Estuary with its off-shore sands, claimed as many victims as the sand bars of the Northern rivers and precipitous rocks of the Yorkshire coast. There another type of lifeboat was evolved, capable of a wider range of action than the Greathead boats. Lowestoft may be said to have had the first effective sailing lifeboat, constructed to the design of Lionel Lukin at the request of the Suffolk Humane Society and built by Barcham in 1807. This became the prototype of the East Coast sailing and pulling lifeboat. It was later developed by Yarmouth's James Beeching when, in 1851, he won the prize awarded by the Duke of Northumberland in a competition to encourage lifeboat design. Soon this was to become the famous Norfolk and Suffolk sailing and pulling lifeboat which could be launched when the local hovellers' beach yawls, which they closely resembled, found conditions impossible.

The Greathead Lifeboat *Tyne* built 1833 and preserved at South Shields. The wide cork fenders and clinker construction marks her out as one of the earlier pulling boats of the North East coast. *R.N.L.I.*

It must be admitted that there were those amongst the longshore communities whose interest in disaster amongst the coal fleet was of a different sort. In a November gale of 1852 the *Maria Elizabeth* was wrecked on Shields' bar and "such of the cargo and stores as have been saved of the elements and the no less destructive fury of the wreckers, are in the custody of Mr. Popplewell, Lloyds agent at Shields and Mr. Turner, collector of customs. A number of wreckers", the report concludes, "have been brought before the magistrates of N. and S. Shields and fined under the Wreck and Salvage Act in sums of from £10 – £1, penalties sadly disproportionate to the offence."

But by the eighteen fifties the length of the coast that colliers sailed had at least some of its terrors ameliorated by the existence of a chain of life-boat stations. When the Swainson Dock at West Hartlepool was opened, "The procession of the opening ceremony was led by 3 lifeboats, fully-manned and equipped, two belonging to the Dock Company and one of the seamen. A finer, hardier set of men could not be seen together than the crews of these boats, who showed clearly their efforts would not be wanting to save life and property whenever required." The community life of the mid-Victorian ports, large and small, found an especial place for the lifeboat movement, providing financial support for boats, building lifeboat houses and providing modest rewards for the crews and helpers. Sometimes local resources alone were not enough as at Tynemouth and later Whitby. In 1861 Whitby decided "to place their lifeboat establishment under the management of National Lifeboat Institute and to accept an offer of the best 1st class lifeboat". Many of these N.L.I. boats were built at Limehouse on the Thames, by Forrest and designed by Peake and paid for by donations sent from inland towns. Over 220 were built to a standard design.

The epic rescues made along the North East coast by these boats and their volunteer crews is typified by the gallant work of the Bridlington station in 1871. It typified, besides, some of the difficulties of using a lifeboat of standard design in a local situation with its especial problems; a trouble which re-occurred throughout the century. An account of the episode given here by John King, coxswain of the present Bridlington lifeboat, illustrates admirably the bravery and frustrations inherent in lifesaving at sea in the days of oar and sail. It must stand for many others, hardly less heroic.

"The winter of 1870/71 had been a particularly severe one and as a result a large number of coal-brigs had been held up and were weather bound in the coal ports of Durham and Northumberland. In early February the weather improved and a great convoy of colliers left bound for the London river and the continent.

By the late afternoon of the 9th February the fleet had reached Flamborough Head and in fact the breeze had died so that the colliers were becalmed right across the Bay. One has to remember, before elaborating on the tragedy of the following day, that the colliers were generally under manned, overloaded and very poorly maintained, with rotten hulls and poor cordage and sails.

The wind began to increase in the early hours of the morning of the 10th February and by day-break a full south-easterly gale was blowing and still freshening. We still get a similar weather situation develop today, usually after three days of hard frost, the frost will break and immediately the wind starts to freshen and always from the South East. Within two hours the wind can increase from almost a calm to force 8 – 9+; it is in these circumstances that we have to launch to escort our local fishing fleet which may well have been caught at sea. If and when the wind does freshen from this direction, Bridlington Bay becomes a true leeshore and dangerous. In any other wind direction the Bay offers full protection. The land itself froms a wide arc and gives protection from the South through the west to N.E. and the off laying Smithic Sands gives shelter in an Easterly. The protection the Bay offers is so good that Bridlington Bay is known as the 'Bay of Safety' and in fact is the only protected safe anchorage on the East Coast between the N.E. Coal ports and the Thames. Even today, in a northerly gale ships (power driven of course!) still seek the shelter of the Bay – particularly, (in passing) the modern colliers, because even though power driven they are, of course, in ballast when steaming North: if a Northerly gale is blowing they make poor progress with their screws half out of the water.

However to return to 1871. There were at Bridlington two life-boats; the official R.N.L.I. boat the *Robert Whitworth* and a smaller privately sponsored boat the *Harbinger.* She had been built in response to the fisherman's criticism of the larger National Boat, as the *Robert Whitworth* was known. They felt she was too big and much too heavy for the shoal water conditions.

However the National boat was launched shortly after day break, and immediately saved four men from one vessel and then pulled out "to the furthest away of the distressed vessels intending to relieve the furthest first." Twelve men were landed from two more vessels. Then for two hours the lifeboat tried to reach a vessel in trouble near the harbour. Before the lifeboat could assist the vessel capsized and her crew drowned. The *Robert Whitworth* had to return to harbour, her exhausted crew had to be lifted from their thwarts and the boat taken out of

The scene at Hartlepool on the morning of 9th February 1861. The morning's gale had seen the rescue of at least 120 seamen by the two Hartlepool lifeboats and one of the National Life-boat Institution's craft, stationed at Seaton Carew. *The Illustrated London News*

service. The local condemnation of the boat was proved right, the boat was too heavy to pull for long periods in bad conditions.

But vessels were still running into trouble. The onlookers sensed that the colliers felt their only hope, as it was impossible for these unweatherly vessels to claw off the lee shore, and the wind was now hurricane force, was deliberately to run their vessels aground and hope the crews could get through the breakers to the shore. Two vessels attempted to enter harbour, but heavy breaking seas were sweeping across the entrance, and both were driven into the harbour wall, breaking up and their crews disappearing into the raging seas.

Now the little life-boat *Harbinger* was manned, she was kept in davits on the North Pier. When the crew found they could not get out of the harbour through the heavy breakers, volunteers lifted her from the water and carried her nearly a mile across the North beach where there was a little more shelter. The *Harbinger* proved to be a fine little boat, being light and easy to handle. She is reported to have saved the crews of five ships and landed each ship's company in turn. Any crew-man of the life-boat who was too exhausted to continue was replaced, and so in the late afternoon David Purdon, the actual builder of the *Harbinger,* took his place in the boat. Robert Hopper, one of the three survivors, described at the inquest what then happened to the *Harbinger.* 'We were about to take an old man off one vessel. He was in the rigging, the rest of the crew having abandoned ship and vanished. A heavy sea came and broke over us as we were in a trough. The boat turned over and we were all thrown out. Two held onto the boat and climbed onto the bottom. I tied myself on with my scarf or I would have been washed away.' There were only these three survivors. Amongst the six lifeboatmen drowned was David Purdon.

There was now no life-boat left, the *Robert Whitworth* being out of service as there were no crew fit enough to man her.

But ships were still being wrecked. A three master disappeared off Flamborough Head and a crew of at least 14 hands perished. Three more ships foundered in the dark. But by midnight the tide had ebbed and men could get ashore and the wind began to die. The next day the beach was littered with debris. Exactly how many men died is not known but local R.N.L.I. records indicated that 70 men were known to have died, as their bodies were recovered; seventy vessels lost. On the Yorkshire coast as a whole 100 vessels were lost that day."

The effectiveness of the lifeboats was considerably increased by the assistance of steam tugs. When available they would tow the lifeboat up-wind of the wreck and then, with the coxswain judging the situation as best he might, he would cast off from the tugs tow-line and drop down-wind. The next stage, and almost always the most dangerous was when, with the kedge anchor down and the cable paid out to control the rate of approach, the lifeboat came alongside the wreck in an

After her crew had been rescued by lifeboat the *Sarah Ann* of South Shields stranded at Tynemouth, one of the victims of the great gale of 9 February 1861. Fifty vessels went ashore between South Shields and Whitby. Of these over thirty were a total loss. *The Illustrated London News*

attempt to take aboard survivors. When this was completed the rescuers would either try to make contact with the tug once again or, should this be impossible, under a scrap of canvas, roar down wind and search for refuge from the gale in a harbour further along the coast, fervently trusting that when it was reached it would be possible to enter.

The paddle tugs of the Tyne which assisted the lifeboats during the nineteenth century were as much a part of the robust character of the North East Coast as the collier ships themselves. The tugs allowed the fleet to leave the river despite a head wind or calm. In weather conditions that, while tolerable in the open sea, made the entrance of the three rivers dangerous the tugs would also tow the colliers into safe water. The ancient tradition of laying up in the dark winter months was therefore gradually abandoned. Towing-out by using the ship's own boats had been regularly employed in calms, or to assist an unweatherly vessel in going-about in an attempt to make the best of baffling summer calms. No doubt the numerous crews of the eighteenth century colliers earned their keep to some extent in these situations but it was an expedient that had a limited use. The regular availability of steam tugs brought a new tempo to the coal trade and with it the opportunity to share some of the coal-factors' profits in the more lucrative winter sales when coal prices soared.

The tugs combined the new form of motive power with an ancient form of wooden-hull construction, a combination of old and new which was mirrored in the outlook and style of their masters and crews. An attempt to introduce a steam ferry on the Tyne produced, inadvertently, the first tug. Unlike the Thames, where ferries were growing in popularity immediately after the Napoleonic wars, the ferry-venture failed. On the Thames there was a traditional hoy-trade which was exploited successfully and combined with packet work. On the Tyne there was no holiday-making to a Geordie equivalent of the Cockneys' Margate or Gravesend. The North East was having no truck with a water-transport that offered three-piece bands, harlequin painted funnels and served light refreshments. So, although a small wooden steamer, later named the *Perseverance,* made history at the Ascension Day jollifications on the Tyne in 1814 by being the first steam boat to appear on an English waterway the enterprise remained unfulfilled. When it was used to initiate a ferry service lack of patronage made the venture a failure. However, upon re-commissioning in 1818, after a period of being laid up, the passenger-work was combined with the towing of sailing vessels. The *Friends' Adventure,* a trader to Hull with general cargo, has the distinction of being the first vessel to be towed across Tyne bar. Initially, Joseph Price who owned the pioneer *Perseverance* and another early steamer the *Eagle,* received little encouragement from the collier owners, but the owners and wharfingers involved with the passage boats and hoy-trade, who had a time-table of sailings to maintain, quickly seized upon the advantages steam-towing conferred. Gradually the range of vessels taking advantage of the innovation progressively widened, stimulated no doubt by the improved power and reliability of the tugs themselves. In 1838 the unappreciated Joseph Price stated that "previously no vessel larger than 240 tons register ever attempted to come up to Newcastle; after the introduction of the towing system vessels of 400 tons register could be brought up; and vessels that previously averaged only eight voyages in the year between Tyne and the Thames were enabled to average thirteen voyages, thereby keeping the Coal Market in the Metropolis and elsewhere, regularly supplied, and preventing those great fluctuations in prices which in former times had such a serious effect in increasing the misery of the poor."

Within a few years the steam-paddle tugs had become an integral part of the organisation of the coal-trade and they were to be found handling colliers and other shipping in every port on the coast, beginning with Blyth. The navigational problems of Blyth were such that the advantages of steam towage were very rapidly appreciated by the shipowners of the port and its colliery owners inland. A contemporary account paints the picture well.

"The (steam) boat arrived in the harbour between 7 and 8 in the morning of the 18th June 1819. In the forenoon the *Resolution* Captain T. Hogg coal-laden was towed as far as the outer beacon, to the great satisfaction of a number of spectators. A brig and a sloop both laden were towed away to sea in the same style. The steamer then returned to the quay when a party of shipowners etc. went on board and spent the afternoon in great conviviality."

The skippers of the remaining 61 vessels lying windbound in the harbour must have sucked their pipes and realized that despite their scepticism a new order of things was at hand. In 1821 there was a tug working on the Wear and by 1838 twelve were towing colliers away from the newly built quays and staithes on the banks of the rapidly developing Tees. Perhaps because of the insatiable appetite of their boilers for good quality steam-coal, paddle-tugs were relatively slow to find popularity on the Thames. In the thirties a ton of coal could be bought on the Tyne for some 11 shillings while it cost 20 shillings from a Bermondsey wharf. The first tug to arrive on the Thames, the *Lady Dundas* did not come from the Tyne until 1832, to be followed shortly after by the *Wear* and both proved of limited use. It was not until twenty years later that they became an established part of the commercial life of the river.

Harton Staithes, South Shields, about 1880. The paddle-tug *Stag*, fitted out for trawling, lies ahead of the barquentine *Aspasia*, loading from a coal drop. A wagon is about to be lowered by the drop. The *Aspasia* was launched at Bideford in 1870 and owned at Guernsey. *Bowes Museum, Barnard Castle*

The *Wear* and *Lady Dundas* were clinker-built, like all the early paddlers. Strangely this most primitive form of boat construction was used for a new ship-type and the tugs reached a size which made them the largest vessels ever built in this way since the later Middle Ages. Iron and steel hulls were only introduced intermittently; the first iron one in 1840, only becoming common much later, with the development of the screw-tug and they were never common on the Tyne until recent times. The paddlers were built at improvised ship-yards on the river-side and averaged between 80 and 90 feet in length. The oak keel was laid, in the time honoured manner, on blocks and marked off into five equal parts. The forward of these divisions and the after-most one were known as the 'huddings', a word inherited from the keelmen. A rabbet, or deep groove, was cut into the central three fifths, to be gradually run out at each extremity. These were cut to accept the two lower strakes or planks of the outer skin which had to be steamed in an iron chest and then twisted into position to lie snugly against the keel at the bow and stern as they emerged from the rabbet. Seven strakes on either side of the keel were required to complete the flat bottom of the hull. Then the sides were built up of larch, 22 strakes on each side and copper fastened, using the same technique of construction as that employed when building a rowing dinghy. The floor timbers to give strength to the wooden shell were then fitted, made from steamed oak and crossing the keel. Between each of these frames was a futtock, rising up the side of the hull for some nine strakes, but notched into the keel and not crossing it. The floats of the iron paddle wheels were of elm, to withstand the rot which constant emergence might cause. The sponsons, which supported and protected the iron paddle wheels were wooden and massively built.

The engines fitted to the tugs were of the side-lever type, of between 25 and 35 nominal h.p. and were a modified beam engine. Unsophisticated by later standards they answered very well, providing an ability to go astern easily and being surprisingly adaptable to manoeuvring the vessel. By operating the main slide-valves in conjunction with an eccentric sheave on the main paddle shaft with a simple device, the engine could go ahead or astern by depressing a pedal on the engine platform. An expansion slide-valve was also fitted and used for economical long steady running. Piston and cylinder were lubricated by nothing more scientifically produced than melted tallow, while joints in the condensing equipment were not machined to fit but encouraged to rust into an airtight joint. The large quantities of low pressure-steam which were needed for the vertical cylinder were provided by two boilers, never lagged and so the top-sides of the wooden hull suffered accordingly. Salt water was used and as might be imagined the deposition of salt in the boilers was phenomenal, it could reach five times the density of sea water. This could cause disaster by causing collapse of the tubes and a simple test was used to ascertain whether a dangerous level had been reached. A bucket of water was drawn off and if it supported a potato a change was necessary; if it sank no

Off West Hartlepool a paddle-tug is struggling to tow a brigantine into deeper water.

Rev. B. Hervey-Jellie

immediate action was required. Changing the water in the boiler required hard work at the hand pump. Leaks from the tubes were frequent and temporary patches were applied utilising a nut, bolt and washer, liberally smeared with red-lead and tar. Sometimes, it must be admitted, such cheerful expedients failed. In 1872 the *Willing Mind,* a tug built at South Shields in 1858 spectacularly foundered and the tersely expressed finds of the enquiry into the loss were "Explosion caused by neglect of the engines and want of water in the boilers."

The tugs gradually tended to increase in size but never exceeded 95 feet, still clinker built, and the larger ones were used to go 'seeking', searching out at sea for homeward bounders in need of a tow, and for salvage work along the coast. The smaller, less powerful, usually contented themselves with river work, but would assist brigs out over the bar for which the charge was a modest fifteen or twenty shillings. They would then anchor and await the possibility of a chance to tow in a wind-bound vessel. Despite the fact that the heavy tow-rope made the helmsman's position in the stern uncomfortable, not to say dangerous, all the earlier tugs were tiller steered and only later was it combined with a traditional wheel mounted on the forward side of the engine-house on deck. Tillers produced a quick response and were unbeatable for steering in congested docks. Finally the wheel was mounted and operated from the centre of the bridge connecting the two sponsons. Tall funnels, which gave a hearty draught, were built for lowering in order to pass under the Swing Bridge at Newcastle after it was built in 1876. Like the sailing coasters many of the Tyneside paddlers were owned in sixty-fourth shares, frequently distributed upon a family basis. Towing and salvage joint stock companies emerged with the construction of much more expensive iron tugs, fitted with powerful twin disconnecting side-lever engines; but that eventful era belongs to the years beyond our present story.

The early clinker-built paddlers were constructed with a distinctive forward raking stem and set a modest spread of canvas; some even sported a bowsprit, surely an unnecessary encumbrance in the towing trade. The rake of the bow gradually became less, but never became vertical. While strictly outside the bounds set by my title the story of the paddlers cannot be left without briefly mentioning another venture in which they prospered. In 1877 William Purdy, a native of North Shields, being a part owner of the old wooden steam-tug *Messenger* conceived the idea that, if towing sailing craft was becoming less remunerative and more competitive, he might try his hand at steam-trawling. Although tentatively experimented with elsewhere, steam-trawling was hitherto unknown on the North East coast. Purchasing a suitable net at Grimsby, a trawl-beam at Sunderland, trawl-heads at North Shields, he then fitted a derrick aft to handle the fishing gear himself and set out. Although one would have thought that handling a beam trawl over the clinker-planked side of a paddle-tug a self-imposed penance he sailed and caught not only fish, but raised a welcome five pounds by towing a brig into harbour! After that, success came quickly and soon there was not an idle tug left at the moorings as his example was quickly imitated, not only at Shields, but all along the coast between Middlesbrough and Aberdeen.

The last surviving wooden steam paddle-tug was not to be found on the Tyne, Wear or Tees. What must have been the last clinker built example was the *United Service,* launched at North Shields in 1871. She was sold to Yarmouth owners the following year and was hard at work at the port until the beginning of the Second World War. She was broken up in 1942.

The off-shore navigational problems of the collier-masters were simplified to only a meagre extent by the guiding lights which marked the worst hazards. This was despite the fact that the enormous volume of the coal-trade contributed mightily through the levying of dues to their maintenance. Until the nineteenth century the lighting of the coast was so inadequate and conducted with such a cynical disregard for the service that it provided for mariners that Samuel Pepys, in his capacity as Master of Trinity House, remarked upon "the evil of having lights raised by and for the profit of private men, and not for the good of the public or the relief of the poor seamen." It was an evil that was to continue for more than a century and a half after Pepys committed his typically kind-hearted plea to paper. Until the Reformation the lights that were provided were lit by the Church, no doubt originally from a conviction of Christian charity, but in the course of centuries they had attached to them the right to levy dues upon passing shipping. Tynemouth's lights traditionally levied a toll of four pence for each English ship and twelve pence for each foreigner entering the Tyne. As in so many other cases this lucrative right was farmed out and providing that a coal fire was maintained and the iron-work of the chauffer kept in good repair no further questions were asked. Light dues, always increasing as shipping grew numerically and in tonnage, were collected by arrangement with the Customs at the Customs' House where vessels received their clearance. In 1665 a Colonel Villiers, a name redolent of patronage and perquisites in Stuart times, had secured the control and monopoly of the Castle light in his capacity of Tynemouth's Castle Captain. He had further increased the light dues threefold by negotiation with the ship-masters of Sunderland and Shields and, in exchange for an annual fee to the Crown of 20 marks, he was to have for himself and his heirs the lighthouse tolls. True, he constructed a new stone lighthouse, but shrewdly economical as ever, quarried the stone and lead needed from the adjoining medieval priory. Finally, in 1674 he built on ground he had demanded as a gift "a home at the cost of £1000 the better to enable him and his kinsmen to maintain the light".

The tower lasted until 1774-5 when it was rebuilt. Some improvement was at last made in 1802 when the inadequate illumination of the coal fire, for which there was no way of giving a characteristic, was replaced by a copper lantern, lit by candle and revolving by means of a clockwork mechanism. This was installed by Captain G. Fowke R.N., the final inheritor of Villiers' original modest investment. By this time London Trinity House had been laboriously and expensively trying to recover the ground lost round the coast by the Stuarts and in 1840 they negotiated the re-purchase of the light and all rights to its dues upon shipping for £125,687, an enormous sum in those days. The old light did not last much longer for it was replaced by the one on St Mary's Island in the mouth of the Tyne in 1898, a change made possible by more modern illuminants.

This pattern of the ownership of a light transferring from church to municipality, or even Trinity House itself, then to be held in private hands and finally repurchased at enormous expense by Trinity House in the nineteenth century, could be paralleled all along the colliers' coast. At Winterton, Lowestoft, Orford, where even the post of lighthouse keeper was used as a bribe at election time, and at Harwich, the story was the same. In 1648 Gerrard Gore, a wealthy coal merchant of Newcastle, secured the lease of the Orford Lights. Was his interest one wonders in the dues or to ensure the safety of the coal-fleet as it stumbled southwards in a northerly gale? At all of them vast profit accrued at a very modest cost by burning flickering coal-fires in tumble-down towers tended by poorly paid keepers who had little equipment save a pair of bellows. Happily some exceptions did exist; on the Blyth the lights and beacons were the responsibility of the owners of Cowpen Colliery and the local pilots — a sensible arrangement. Cromer was provided with a light in 1674 and then it was renewed by the efforts of the coal-masters of Ipswich, Newcastle and Yarmouth. Two lighthouses marking the changing entrance to the Humber were provided by a London merchant, Justmian Angell and maintained by a levy on shipping using the river. The Humber, apparently such an obvious haven of shelter on the map, has strong tides and a low featureless coast so that its dangers were avoided in extreme conditions and the lights only partly improved the situation. A collier, grounding on the sands could actually find itself upset as the fierce ebbing tide scoured out the sand beneath its bilge.

At North Shields the first lit beacons to be established there in the post-Reformation age were established by Newcastle's Trinity House and they were kept securely under their jurisdiction. Henry VIII granted to the brethren of the Guild a licence in 1539 to build two towers which were to combine defence with the guiding of mariners, on the north side of 'le Shelys' only a cannon shot from the Durham shore. Their construction was encouraged by a Royal instruction permitting the use of material looted from the Dominican monastery in Newcastle which was brought by keels down to a site on the left bank of the Pow

The brigantine *Mary and Agnes* ashore at Whitby in an autumn gale, 1885. She had been built at Sunderland in 1842 and was owned at Scarborough at the time of her loss. *Sutcliffe Gallery*

burn. The High Light was constructed so that it and the Low Light were in direct line with the then deep-water channel used by colliers at the mouth of the Tyne. Contrary to the conventional method of illumination these towers held lanterns lit by candles, suitably shielded by glass. The effectiveness of the lights were subject to the same problem of any pairs marking a sandy entrance channel; the channel may vary over a period of years, leaving them useless if not dangerous for navigation. This occurred at North Shields and in 1658 they were demolished and replaced by wooden towers which could be cheaply moved and re-erected. However, in 1672 the Trinity House of Newcastle decided to revert to a stone tower again and they built this time within the walls of Prior's Fort. Every English ship which entered the Tyne was levied twelve pence and every foreigner three shillings. A high-house was built in 1727 and this is still standing, rather sadly alone for many years amidst twentieth century urban decay. Candles were still the only form of illumination in 1727 although it was proudly pointed out by Newcastle that they had increased them from two to three. Again in 1805 the old lights were condemned, convicted of no longer marking the channel accurately. New lights were built in 1806-8 on another site and these still stand, while the old towers were, appropriately enough, converted into Alms' Houses.

The life of an eighteenth century lighthouse-keeper was far from the lonely and monotonous routine we associate with it today. Like the ships they served they too suffered attacks from the enemy. In 1701 John Hooke, in charge of the Orford light wrote despairingly to his master, the Marquis of Hertford, "There have been a privateer a shoote at the lithouse and have broke all the glase and stoll all the Ierons belonging to the furnes and they stoll the mens beds and I wonder they did not burn the house and if there be not sum speaddy care taken to prevent the same I shall not git eany men to louck after it for the say will begg there Bread before they will com there eany moore."

The disasters which drove vessels to within sight and sound of the shore provided the most heart-rending tragedies. Here, in savage broken water too shallow to permit the would-be rescuers from floating out a boat the helpless watchers could only see the crew, lashed in the rigging, hoping the tide would ebb and give them some chance before the vessel broke up. Beginning with Captain Manby of Great Yarmouth, who had witnessed such scenes, there were various attempts to develop a line-throwing projectile to establish communication between the shore and the ship. Manby in the eighteen hundreds favoured a mortar, others used rockets, but the object was to throw a line over the wreck and then eventually a rope substantial enough to use a breeches-buoy. Manby's first success was achieved when he rescued the crew of the brig *Elizabeth* in 1808. However, the practical problems were far from solved by Manby or other inventors. By the mid-century this method of rescue had earned official support and had been sufficiently improved for the Coast Guards, originally formed to suppress smuggling, to be equipped with rocket apparatus to ring the coast. They were supported by voluntary groups, and the first Volunteer Life Brigade was founded at Tynemouth in 1864, an example soon followed by South Shields, Cullercoats and Sunderland.

The worst dangers of hidden shifting sands, located off a low-lying coast, can only be marked by buoys and light-ships which may be re-aligned without trouble. As early as 1606, Trinity House of Deptford issued an order to its officers at Newcastle, Great Yarmouth, Hull, Lynn and Boston to levy a tax on shipping using these ports for the maintenance of buoys and beacons between Lowestoft and Winterton. But this early enthusiasm waned. It was left to private initiative to prove that 'floating lights' were a practical proposition and the first lightship to be built was placed to mark the Nore Sand in the middle of the Thames Estuary, in 1732. Trinity House strongly objected to such an innovation and its further infringement of their monopoly. The lightship carried two large 'lanthornes' on the extremities of a single yard which was lowered to deck level for servicing the lamps. Although it was not taken over by Trinity House until 1794, its obvious success spurred officialdom to further, if reluctant effort, and their first lightship was anchored at the Dudgeon station in 1736. Just as the Nore light provided a guide for the converging shipping in the Thames mouth so the Dudgeon, placed at the especial plea of the collier-master, lit their way across the waste of water between the Spurn and the high land of the north Norfolk coast and warned of the dangers of the Wash, graveyard of so many colliers. The lightships, as they multiplied with the Happisburgh in 1789, the Newarp in 1790, one at the entrance to Yarmouth Roads in 1843 and the Cork, near Harwich in 1844, did more than provide points for navigational reference. They forged another link in the chain of assistance available to vessels in distress by firing a warning gun to alert the lifeboat or to warn a collier that it was standing into danger.

The longshoremen and salvage cutters earned their living from the misfortunes of the collier-fleet. They gave assistance to ships in trouble by taking out extra hands for working the pumps, provided specialist knowledge and skill to extricate a collier in danger. Ground tackle was ferried out to brigs which had 'cut and run', and pilots were put aboard. It was work which offered temptations. There could be collusion between longshoremen and captains to make spurious salvage claims and to share the payments made by the insurers, boatmen could convince worried masters that they were in imminent danger when none actually existed and exorbitant charges might be made for anchors which had been swept up from the sea-bed at little cost. Nevertheless, they undoubtedly performed sterling work by rescuing life from wrecked vessels when there was little hope of monetary reward from salvage, just as it was not unknown for lifeboat crew to make claims upon insurers after their first duty of lifesaving was done and property had been

One of the early Tyne paddle-tugs, the *Flying Childers*, built at North Shields in 1857.
Omell Gallery, London

saved by their exertions. A summer occupation of the Yarmouth and Lowestoft hovellers was to retrieve the cargoes of wrecked colliers that had been the victims of winter gales. They devised dredges similar to the Thames ballast-dredgers' simple machinery and brought up sizeable quantities of coal from the broken hulls lying in shallow water along the edges of the Scroby and Newarp sands.

Rewards were presented by the N.L.I. to hovellers carrying out rescue work in their own boats as when the crew were saved from the brig *Gloucester* of South Shields, lost off Filey in the January gales in 1861. At the same time the *Eliza* of Middlesborough was lost with all hands at Steaton Carew while the brig *Arethusa* of Blyth came to grief on the Cross Sand. She was fortunate, for before breaking up, the crew of eight were rescued by the life-boat. The next week saw six brigs and a schooner, all in the coal trade lost on the North East Coast and at least 30 seamen owed their lives to the work of the lifeboats and their crews. Further down the colliers' coast the men of the brigantine *Atlas* of

Stockton were picked up after they had abandoned their leaking vessel exhausted with days and nights of pumping. They were fortunate enough to be picked up by a Danish schooner who in turn transferred them to a Yarmouth fishing smack. At Yarmouth they found at least a temporary refuge and welcome in the newly built Sailors' Home. Although destitute and far from their home-port all too few victims on the sea were as fortunate as these men. A few years before the grim epitaph to a score or more seamen was recorded in the local newspaper with a paragraph which stated "A laden galleot, a schooner and a brig were observed foundering off Hartlepool and Tees. Names unknown."

North of the Tyne, Nathaniel Crew, Bishop of Durham provided posthumously in 1721 a system of warning and rescue. Administered and maintained by the Crewe Trustees it was centred on Bamburgh Castle and was an elaborate scheme for assisting mariners in distress. It helped to provide for survivors and included the provision of 'premiums' to the first longshoremens' cobles to put off to vessels in danger on the Farne Isles. They also maintained a lifeboat at North Shields. The work of the hovellers in assisting the colliers was recognised by the London Coal Exchange by providing them with brick-built look-out towers at Aldeburgh, on the Suffolk coast. The Coal Exchange also subscribed generously for the provision of life-boats which, in the tradition of the Institution carried their name.

That the efforts of the lifeboats and their supporters was doubly necessary is shown by the situation along the coast a few years after the National Lifeboat Institution received Royal patronage. In 1867 in an account of the casualties it is pointed out that of the 2,513, more than half were represented by "unseaworthy, overladen, or ill-found vessels of the collier class." No fewer than 713 laden colliers and 242 colliers in ballast were helped which gives an opportunity to assess the total size of the fleet involved. Of the total disasters recorded of every type of vessel nearly half occurred on the East Coast. After analysing the age of the vessels lost, the report ends by regretfully stating "that even at the age of twenty-five to thirty it sometimes happens that a ship is so rotten as to fall to pieces immediately on touching the ground, without the slightest opportunity of the crew getting out their boats, or to be saved by life-boats or the rocket apparatus". So much for those who imagine every old wooden vessel was a triumph of unlettered craftsmanship and built to last a century.

The employment of pilots to assist shipping on the North East coast is the oldest of the services to wait upon the coal-trade. Trinity House insisted that pilots were, as part of their qualification, to be freemen of Newcastle and no vessel might be guided over the bar, though she were sinking under the feet of the crew, unless the pilot had this distinction. Blyth and the newer ports had a pilotage system which was less severe in its selection of candidates. The pilots elsewhere on the North East Coast held certificates of authorisation which Trinity House of

London issued upon successful examination of the men concerned, and renewed at intervals.

They used the traditional North East coast coble to reach the craft needing assistance and were paid according to the draught of the vessel. The pilot cobles working from Shields and Sunderland were built with a narrow transom while those from Redcar, which served the shipping entering the Tees, were sharp sterned and had the further distinguishing mark of being painted black instead of the traditional white, red and blue. They varied between 20 and 25 feet overall and in the eighteen forties, when their numbers reached their height, cost £25. A coble was rowed, as well as sailed, with their heavy, characteristic square-loomed oars and carried a tall dipping lugsail and jib. When sailed they needed the addition of up to half a ton of shingle ballast stowed in the bilge in canvas bags. Their number and a white and red pilot flag were painted on the bow indicated they were licensed by the pilotage authority.

A mid-nineteenth century account gives an insight into their activities. "The Tyne pilots go to sea in open boats called cobles" it opens, and continues, "They

The Peake self-righting lifeboat *Florence Nightingale* stationed at Sunderland 1865. She pulled ten oars and was 33 feet O.A. with an 8 foot beam. The high launching carriage enabled her to be used with the maximum chance of success. *R.N.L.I.*

are exceedingly smart little craft, but have no convenience for shelter of for stowing away the smallest quantity of provisions. The men sometimes run as far South as Yarmouth Roads (some 500 miles South of the Tyne) in these small vessels to meet ships, and are indebted to those they meet for a supply of provision which in most instances are freely given". The only shelter aboard for the complement of four or five on these coastwise cruises in search of a ship was a canvas tilt spread over the lowered mast. Cooking was done over a 'fire-pot', held in place over a few stones in the ballast. When a coble was towed it was secured stern-first and the rope never made fast but held by the crew, the risk of an over-turn being too great to permit even a hitch around a thwart.

The cobles were for sea work and for employment offshore, in the river the foy-boat was used by the watermen. They were essentially lightly built rowing boats, with a narrow transom stern and without the deep forefoot, or 'ram' and twin keels of the coble. A dipping lugsail was carried, but without the ballast or deep keel they were difficult to handle under canvas. Some 15 feet in length and 4 feet 6 ins. beam they had two thwarts and a board at the stem and stern. The foy boatmen assisted the sailing ships in mooring, putting the skipper ashore, carrying supplies, and generally served shipping in the Tyne. The Tyne piers were the usual limit of their activities; however, sometimes they picked up their ship, in days gone by, well out to sea, and were towed in, like the cobles, with the tow-rope ready for immediate release. They were clench built of oak and elm and the builders whose work was most respected in later days were Hales and Mitchelson, both firms who worked at South Shields. The name possibly derives from 'foy' , being a corruption of 'fee', the cash remuneration which the foy boatmen earned for their work. The light clinker-built construction and fine run certainly point to a Scandinavian origin of the foy-boat and perhaps the name has an origin in a similar area. At the end of the eighteenth century it was estimated that together pilots and the foymen on the Tyne numbered some five hundred. Unlike the keelmen, the coble-pilots and the collier-brigs, the foy-boats and men survived into the twentieth century assisting the steamers in the same way as they had provided a service for the sailing-vessels, although working on a rota and not, as in the past, racing to be first alongside ships coming in over the bar.

0 1 2 3 4 5 6 7 8 9 10 15 20 25 30 35 40

SAIL PLAN OF COLLIER c 1850.

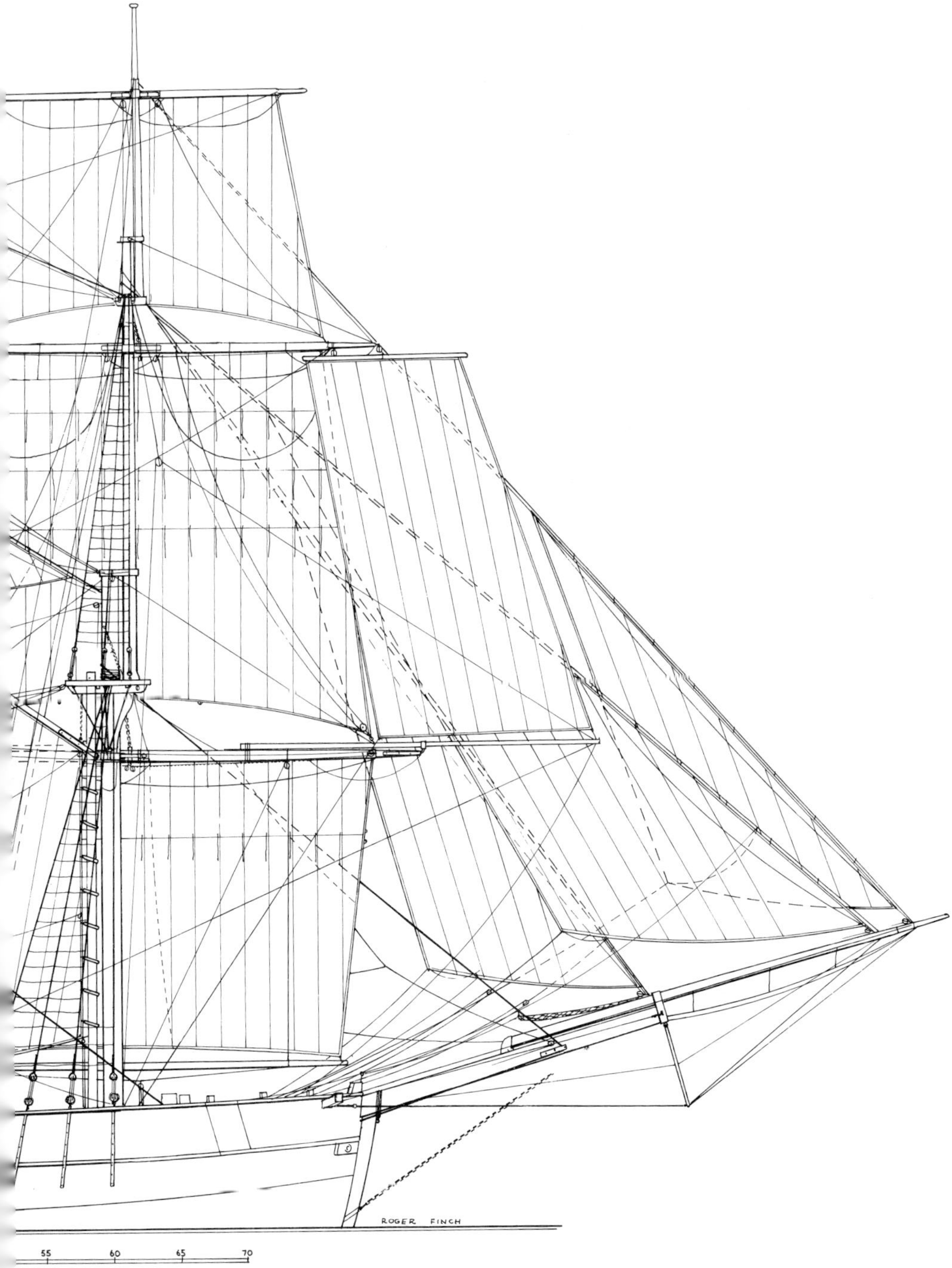

10 Nineteenth Century Expansion

"But most assuredly, nothing that ever swung at the quay-sides of Carthage, or glowed with crusaders' shields above the bays of Syria, could give to any contemporary creature such an idea of the meaning of the word Boat, as may be now gained by any mortal happy enough to behold as much as a Newcastle collier beating against the wind."

John Ruskin

THE coal pits on the Tyne had been using steam-powered pumps in order to drain the workings, at least since 1715. In that year a Newcomen engine had been installed at Tanfield Lea Colliery and their numbers had grown until by 1769 more than 100 were in use. The tentative use of steam locomotives to replace the horse-transport along the wagon-ways leading to the riverside spouts dates from 1808, and William Hedley's "Puffing Billy" was heralding a far-reaching change for the nation when it triumphantly hauled coals to Lemington staithes in 1813. Steam-propulsion ventured afloat on the Tyne when the *Perseverance,* the first steam packet on the river, made her appearance. These were but three examples of a new quickening of development and a swifter inter-action between growing demand and technical innovation. There was a parallel hint of the future effects of a coal-based technology upon the Thameside industries. Some had the most modest of beginnings. Those who saw the discarded brewers' vats in use to contain the newly produced 'coal-gas' which had recently lit Pall Mall and now, it was said, would be available to the houses of the nobility and would also light the new manufactories, could hardly have envisaged the thousands of tons of coal it would consume yearly, within a few decades. Newcastle boasted it had lit a street by gas as early as 1818 thus leading Britain in its use. However until at least the mid-century, the growing complexities of industry, with steam as its prime mover and steadily mounting domestic demand for coal were both served by the humble sailing collier. The amount of coal coming to London at the beginning of the nineteenth-century was rather less than 1 million tons; by the eighteen fifties it was nearly 4 million and all delivered by sail in wooden hulls.

But the steadily mounting consumption slowly forced a change of pattern in the sea-coal trade; a change to contracts and direct purchasing by industries, such as brewing. It slowly eroded the old pattern of masters of ships buying and selling; they became carriers, making such a profit as they could from the transportation of cargoes and allowing others to organise their purchase and sale. The ship-masters were far from regretting the change. In 1832 they published a broad-sheet, enjoining their "Fellow ship-owners . . . to make a stand until we can obtain a fair and reasonably remunerating Freight for the cargoes which our Ships carry, and I would strongly recommend no Ship-Owner to purchase one . . . it is perfectly becoming in us to take the most simple position by letting our vessels 'lay by the walls' until a Freight can be obtained, which will pay us for our hazardous employment and uncertain property . . . "

As in the North the sheer volume of the trade put the severest strain upon the capacity of the Port of London to handle it. The situation was aggravated by the fact that all dutiable goods, and most goods carried a duty, could only be landed at legal quays although with the exception that coal might be discharged at the Coal House at Gravesend. There were enquiries into the working of the Port of London to attempt to find a solution, the first in 1799, when it was stated that "The Colliers, by their repeated Voyages, exceed in Number of Ships and Tonnage, those employed in the Foreign Trade. The Importation of Coals, on an average of Seven Years preceding 1782 was 474,717 Chaldrons; it now amounts to 900,000 Chaldrons per Annum and it will probably increase and call for more room." In 1785 it was stated that as a result of an intervention of the Lord Mayor 130 out of the 220 vessels in the river were quickly unloaded and fuel was once more available. It gives an idea of the size of the fleet, confirmed a few years later, when in a committee report it was said that colliers were the most numerous class of vessel in the river and represented three-fifths of the coastal tonnage entered at the Customs House.

The London river with its hazards, natural and man-made, had an important effect upon the size and rig of the nineteenth-century collier. The earlier estuary charts show no more than a few of the more notorious sandbanks until after the middle of the eighteenth century and are more adapted to interior decoration than navigation. But they reflect the fact that while the channels presented difficulties in maintaining direction to a sailing ship there were no definite limits due to lack of depth of water. The first chart of the Thames is that of Whitworth,

The first shipments of coals leaving the new Sunderland Docks, opened May 1850.

The Illustrated London News

published in 1775, and tells a different story, and enables us to envisage the coal-barks navigational problems; but I doubt if any collier master knew of its existence, let alone possessed one. They employed water-men, licensed by the City of London, to assist them in safely negotiating the hazards ahead once they had passed Gravesend. Trinity House had vested in it the sole rights of dredging for ballast and would, one might have thought, improved the channel for the benefit of the seamen their organisation was intended to protect. However, the ballast dredgers, whose equipment was of a medieval simplicity, kept within the distance which the power of the tide enabled them to traverse, from the great fleet of shipping in the Pool, and to whom they sold their spoil. This critical distance was about eight or ten miles and the lower river, down below Woolwich was therefore unaffected by their, admittedly, meagre efforts. The collier masters complained bitterly that they, Trinity House, were only concerned with winning saleable clean gravel and sand and ignored the shoals of mud which formed and impeded navigation. The Geordie sailor could also have pointed out that the situation in the Pool had actually deteriorated due to silting resulting from the detritus of London's vast growth. In fairness to Trinity House it must be said that their ballast-men were aristocrats on the River's colourful *lumpen proletarat,* earning on average 21 to 22 shillings per week and enjoying the prospect of a pension and sickness benefit, almost unheard of in other trades.

Sailing navigation between Gravesend and the coal wharfs below Shadwell and Billingsgate was conditioned by the strength and direction of the tide, by the bends in the Channel, and by the shoals. The modern yachtsmen with his fin-keeled, sparse and scientifically rigged craft, or even the school-ship captain (both with the button of their electric start diesel within arm's-reach) must look at the hull-form and the apparent tangle of rope aloft in the model of a collier-brig and wonder how the real vessels coped. The lower channel of the Thames was often wide enough for some tacking and here the Bentick-boom helped. This was a light spar to extend the foot of the square foresail; the clews of the sail were secured by chain to the outer ends and at the centre. These came together and led to the deck, through an eyebolt and thence to a band on the foremast, or were set up with a simple purchase. The control of the sail, which did not need the normal tacks and sheets of a square-rigger, was by the fore-bowlines, one attached to each leech of the sail. These led forward to the bowling block on the forestay. When changing tacks, at the cry of "lee ho!" the jib sheets would be eased and the fore-

The nineteenth century sailing collier at her best. The *Fuchsia* 240 tons was built at Sunderland in 1846 and was owned by Potts and Company. *National Maritime Museum, London*

The snow *Middlesboro'* 209 tons, the first built at the town after which she was named. *Peter Barton*

bowline let go as the vessel's head swung round. It made a crucial difference in time and manpower. There was, however, always the tide to fall back upon, indeed it was rarely that a vessel moved without a fair tide. A dependence on the tide for movement, as well as to provide depth of water was the ultimate fact of Thames navigation, indeed of most rivers, particularly the entrance to the Humber, in pre-steamer days.

In these conditions, as in many others encountered in the north-east coal trade, the brig or snow rig came into its own. Between the middle and end of the eighteenth century this rig had tended to become increasingly common in vessels of up to 200 tons, tending to replace the use of the ship-rig, with its three masts, each setting some form of square canvas. There had always been two-masters, doggers, billanders etc., but they were small coasters, usually less than 80 tons. Perhaps because of the war-time restrictions on materials and the press-gangs' predatory visits to the ports it was forced upon owners that hulls as large as the traditional collier working about the North Sea could be traded as effectively, and far more economically rigged, as a snow or brig with her two masts than as a ship or barque with three. By the time E.W. Cooke produced his immensely detailed engravings of the Thames shipping in 1828 the collier is almost always a brig. Prior to the French Wars, beginning in 1793 the traditional collier is a ship. With this change in rigging the collier-fleet lost something of its splendour, but the owners saved a great deal on cordage and in payment and subsistence to the smaller crews needed by the brigs. The master, nevertheless, had a vessel which in most circumstances would be as easy to handle. The crew of a collier in the days of James Cook numbered at least ten, with two or more apprentices. It was possible to sail a brig with six hands. With eight hands, all told, they would consider themselves well provided for. In the eighteen fifties the construction of barques began again, particularly at Sunderland; but these were world-traders of at least 400 tons and while frequently freighted with coal were in effect ocean-going tramps, and had little in common with the bluff-bowed, stubby colliers which Richardson and his ship-mates knew.

An added incentive to change was probably the less easy-going nature of trade as it stabilized after the post war prosperity, when there was a closer examination of over-heads and as convoys no longer restricted the availability of tonnage. There was a realisation that the brigs, which always existed in the coal-fleet, could manage at sea and in the rivers as well as their three-masted competitors. It is worth noting that the famous centenarian *Brotherly Love,* built in 1765 and rigged as a ship, had long since been converted to a brig by the time of her loss off the Yorkshire coast in 1878.

The brig or snow could negotiate the Thames in face of the prevailing South Westerlies, where rounding the Isle of Dogs required an 180° change of course, and the serpentine bends round Blackwall point and Broadness were almost as sharp, by making 'stern boards'. This was a sailing manoeuvre used when tacking in light winds, through a narrow channel, evolved to overcome an inherent inefficiency in the rig. Upon completing the longer leg of a beat against the wind,

Half-model of collier-brig (1840). *National Maritime Museum, London*

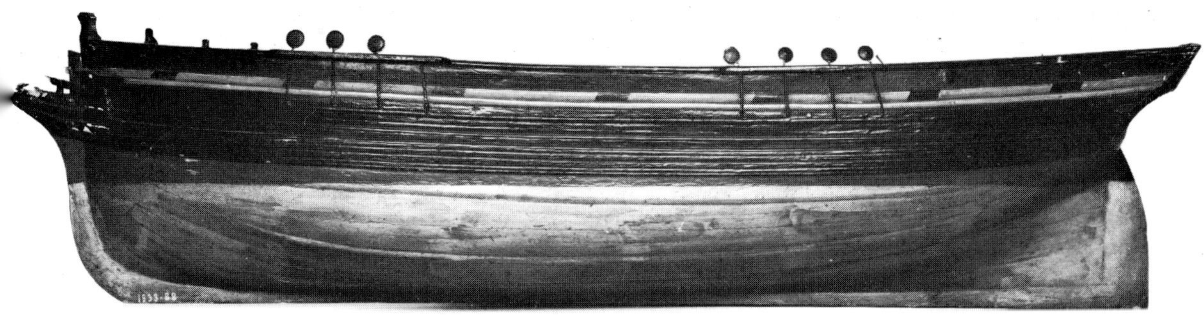

colloquially known as "writing W's after your name", the brig's yards would be swung and with the sails aback, that is to say with the wind blowing into the forward surface, the vessel would be driven stern first across the stream. Despite its unseamanlike appearance it gained ground by the tidal current. This apparently clumsy method of progress reduced the danger of missing stays at the end of the tack and the resulting grounding or collision with the tiers of moored barges which lay just outside the limits of the channel. It took no little skill and nerve to manage this method of progress in company with a score or more companions amidst the hazards provided by drifting lighters, ferries, fishing craft and, as the century wore on, passenger steamers which obeyed no known rules of the road.

Just below Barking Creek was a critical point of shallow water and while the coasters could reach this limit at any stage of the tide, beyond here they had to use tidal water for further ingress. Perhaps this is why Barking and its creek, with its own sea-going fishing fleet until 1854, preserved, before it was overtaken by a flood of surburban bricks and mortar, something of the atmosphere of a coastal port. Just beyond Woolwich was another 'narrow' and this meant that our collier captain must anchor in the depths above the Naval Dockyard there if he could not reach one of the 'holes' as deep patches in the river are known, off Blackwall or further on at Deptford. For many vessels, drawing 16 or 20 feet this represented the inward limit. They had to load and discharge at this point, their cargoes being taken up to the Pool in lighters. It must be remembered that the complex system of docks, now as I write, drawing to the end of their useful life, barely existed at this time and then only for very specialist purposes. The limit of 16 feet draught meant that colliers, in order to reduce the inordinately long period spent unloading to a tolerable minimum, were restricted to a depth of hull less than this. It was a factor which in turn limited their tonnage and so rigged as barks or brigs, ships or snows, a little over 200 tons became the standard size. The coal-factors came habitually to deal in this tonnage and the whole process of discharging, lightering and storage accepted it so that entrenched tradition tended to perpetuate the limitations.

This tonnage accorded well with the draught suitable for loading at the coal drops on the Tyne and Wear, although there the final part of the cargo had to be secured from keels in midstream on the Tyne as the water alongside the staithes was insufficient for a loaded collier. The river entrances were very poor in the Northern ports and any deeper draught than the standard 14 or 15 feet would have courted disaster. There were, of course wide variations in size and perhaps the lower limit was typified by the brig *Permanent,* a regular coal trader, built at Yarmouth in 1790 and of 59 tons reg., by tradition she loaded 100 tons. Another, the *Tyne,* built at Newcastle in 1823, loaded about 114 tons on a 9 feet draught. The *Suffolk,* another miniature, built as a brig in 1818 at Wells of 88 tons reg.

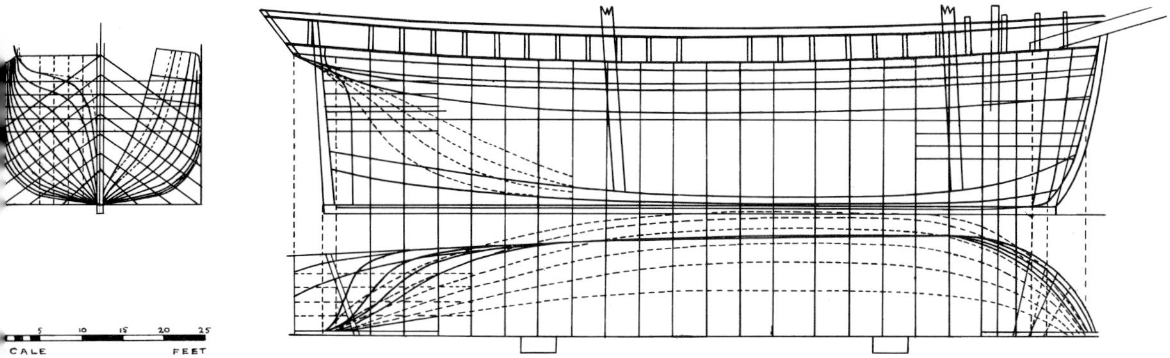

The lines of a collier brig 1853. *Drawn by John Lewis from material supplied by the Science Museum, London*

had a length of 62 feet and a beam of 18 feet 3 inches. Ten years later a one-eighth share in her was sold for a modest £55 by which time, hardly surprisingly, her rig had been simplified to that of a brigantine. The cost, ready for sea, of one of the normal 250 tonners appears to have been approximately £3,500 and depreciation affected the investment fairly steeply.

The single-masted sloops might very nearly reach the tonnage of their square-rigged superiors; like the *Sally,* "by Register 75 tons, and Draught, coal loading nine Feet built at Great Yarmouth in 1776, thoroughly repaired last Summer exceeding well found, and fit for sea." One can perhaps postulate that a draught of nine feet was most appropriate for the creeks, drying harbours and beachwork along the coast.

Ponderous three masters or diminutive sloops, the colliers had reached their final objective when they had dropped their big wooden-stocked anchors in the foul mud of the Pool of London. Here were the concentrations of Legal Quays, dating from early medieval times and the more recently established Sufference Wharves, licensed from time to time and attended by Customs officials who supervised the landing of certain specified dutiable goods. The wharves stretched in picturesque confusion on both banks between London Bridge, below the Tower, as far as Hermitage stairs; an alternation of medieval timbering and gimcrack seventeenth century weather-board, only occasionally punctuated with more substantial Georgian brick and stone. In the stream lay hundreds of craft, colliers from the North East coast predominating, each to his own anchor. The circumstances of the system of restricting the landing of dutiable cargoes at only Legal Quays and Sufference Wharves inevitably concentrated shipping in this short reach of the river. Despite this colourful congestion and the problems it caused

THE FORE MAST, SAILS AND RIGGING OF A COLLIER BRIG. C. 1840.

SAILS

A. Fore topgallant sail.
B. Fore topsail.
C. Fore mainsail.
D. Spencer.
E. Port topmast—studding-sail. (Set forward or abaft topsail according to the point of sailing.)

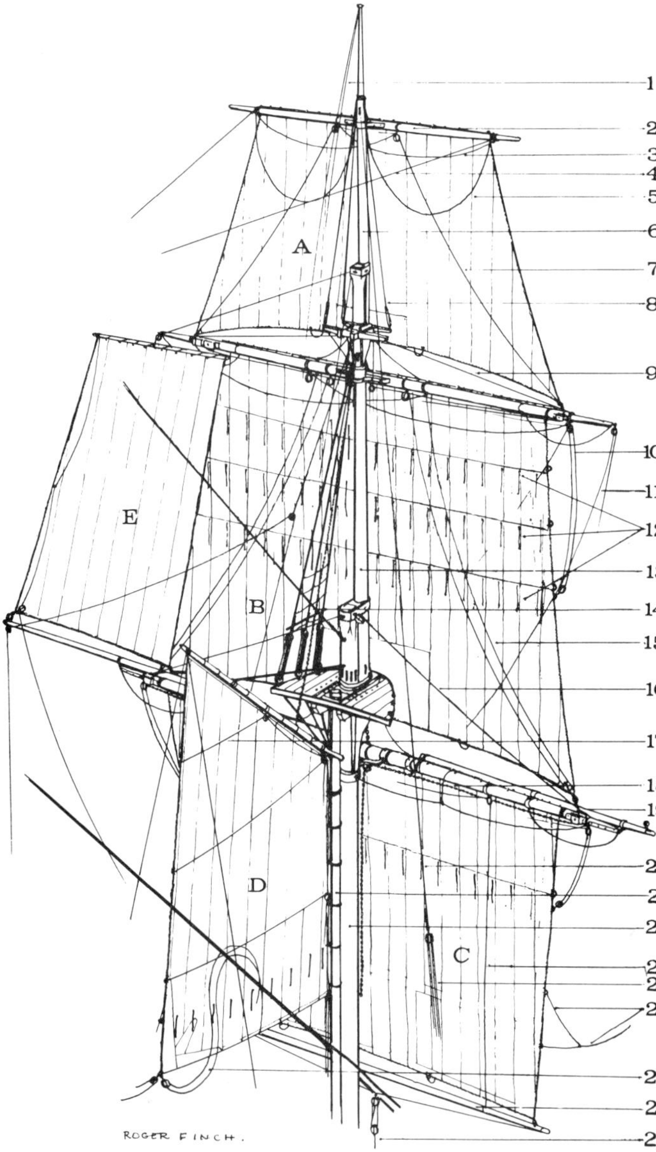

MAST, SPARS AND RIGGING

1 Flag halliyards.
2 Fore-topgallant yard.
3 Foot rope.
4 Fore-topgallant brace.
5 Fore-topgallant yard lift.
6 Fore-topgallant mast.
7 Fore-topgallant clew-lines.
8 Fore-topgallant-mast shrouds.
9 Fore-topsail lift.
10 Yard-arm reefing tackle.
11 Studding-sail halliyard.
12 Reef-points.
13 Fore-topmast.
14 Fore-topmast shrouds.
15 Fore-topsail clew-lines.
16 Fore-mainyard lift.
17 Spencer upper-brails.
18 Chain fore-topsail sheets.
19 Starboard fore-studding sail yard.
20 Fore-topsail halliyard.
21 Trysail mast.
22 Fore-mast.
23 Bentick boom halliyard.
24 Fore topsail halliyard purchase.
25 Fore-bowline and bridle.
26 Spencer sheets, double.
27 Bentick-boom.
28 Bentick-boom sheet.

river navigation it was not until 1836 that permanent moorings were laid for them; these were 11 and 12 ton stones, sunk to the depth of 12 feet, to which the colliers tied up. Hitherto they had anchored with their rope cables, of doubtful quality which often parted, allowing them to break adrift as they swung with the tide and causing further chaos amongst the welter of attendant lighters, barges and river steamers which threaded their way through the grimy fleet.

Having found a berth the master of the collier then supervised the stow of the sails seeing they had a "full bunt and lean yard arms", some were actually unbent from the yards and put below, and then ensured that the gear and appearance of his vessel would not draw unfavourable comment from his fellow captains ranged around him. Making sure that his shore-going clothes would command suitable respect from the factors at the Coal Exchange the captain was rowed ashore in the waterman's skiff which had been towed astern from Woolwich while its owner acted as pilot and extra hand. In the master's lap was a leather satchel containing the 'crocket' from the Customs' House at Newcastle and a certificate from the fitter on the Tyne showing the name of the ship-master and his ship, the port of registration, date of loading, quantity and name of the coals she carried, so that the quality of the cargo might be assessed, and finally the price paid on the Tyne. When the Customs' clearance was made good by the factor, who had the capital to pay the duty, he would arrange for the sale of the coals and for the meter, under whose surveillance discharging was done and who measured out the cargo, to go aboard. The factor arranged for lighters to be brought alongside and these often acted as floating warehouses, further constricting passage on the river. The papers which the captain held were sometimes sent ahead by post-boy from Gravesend, Leigh Roads, at the bottom of Sea Reach, or even further along the coast, in order to secure a 'forward sale'. It was known for them to be dispatched by a fast-sailing hatch-boat from the Nore, or even over-land from Yarmouth. This was objected to by the meters as giving an unfair priority and they insisted that ships should be worked in the order of their arrival. The meter was victualled by the ship-owner while at work and much of the traditional humour of the coal-fleet concerned the expensive tastes of coal-meters and the tendency for ship-masters to indulge them. But the masters had their perquisites as well. Going ashore with the skipper, contemporary letters use this title for the master without hint of patronage, in the waterman's skiff might well be a passenger or two whose modest passage-money went to the master, for the stage-coach fare from the North East discouraged all but the wealthy from using them. The Bewick brothers, whose engravings enliven these pages, made their first journey to London as passengers aboard a collier. Along with the official documents in his satchel would be a list of commissions domestic and modestly commercial, that he had undertaken; the purchase of lengths of cloth, books, raw material for the pharmacist, wine and a dozen items which the metropolis provided and yet the

The snow *Triton* of Sunderland, built there in 1824, 241 tons. Dimensions 87 ft. 3½ ins. x 25 ft. 6½ ins. x 15 ft. 7 in. A collier and Baltic trader. *Parker Gallery, London*

provinces saw but rarely. He might also be undertaking a little private speculation in fish, keg-butter or Hartley glass-ware.

The process of unloading was a monopoly of a group of 'undertakers', also contracted by the factor, who supplied labour and equipment. They had organised themselves into a 'society' and effectively controlled the discharging of the colliers in the Pool, despite well-meaning Parliamentary efforts to break their grip. In the manner of the time their employees were enrolled for work and paid at a public house where only those who drank freely at inflated prices could hope to be selected for employment on the colliers.

The older method of employment was for a gang of 16 coal-heavers, under a foreman, to shovel the coal up from the hold by a series of stages, the wooden stages being temporarily erected by the gang themselves. After the Restoration an unusual source of recruitment to these groups had been the Grenadier Guards, who were granted the right to act as coal-heavers in their spare time in order to supplement their pay! From the deck of the ship the coal was measured into wooden vats and thence tipped into the lighters alongside. The inevitable breakage that resulted from the process, known as degradation, was all to the advantage of the ship as the vat measured volume, not weight. The coal-meters deputy 'topped' the vats and recorded the final quantity, for the King's and City tax were based upon this measurement. Much extra expense was incurred by the meters working office hours and as no unloading could take place without their presence, merchants could do little to hurry discharging in times of shortage. The lightermen were notorious for their ability to fail in clearing the final few chaldrons from a ship and having to be bribed to finish her and to save the vessel another unprofitable day in the Pool.

The lighters, when they had loaded their 21 Winchester chaldrons, and a buyer was bound to supply craft which would clear at least this amount in one day, moved up stream with the tide. Many went to the North Shore, between London Bridge and Hungerford which was taken up with coal wharves and the first gasworks, before the construction of the Victoria Embankment. A typical collier might take six to eight days to unload, providing no time was wasted waiting for lighters, meters and heavers, but in practise ships were kept swinging at their

The little schooner *Bee* discharging coal in the Watermill Dock, Walton, Essex, brought from Seaham, Durham. She returned to the North with bagged flour and eventually was lost in an easterly gale, driven ashore at Withernsea, Yorkshire.

anchors from eight to twelve days. The dependence upon suitable weather for shipping to leave the Tyne and again to move beyond Yarmouth Roads tended to bunch the colliers together and there were sometimes as many as 300 in the river together. The strain these huge fleets placed upon such an archaic system of discharging can be imagined. Lighters, waiting for vacant and expensive storage-space ashore into which to discharge, blocked the channel. The river stank from untreated sewage, coal-dust filled the air, blackened masts and yards pierced the haze above which the golden cross of St. Paul's improbably rose. The clank-clank of ships' pumps told of vessels who, even in the security of a river-anchorage, had to work at frequent intervals to keep themselves afloat and the wooden hulls ground and creaked together as individual brigs sought to extricate themselves from the tiers and start on the long slog home. Now that the fish are back in the Thames and St Paul's stands clear and white, alas the ships have gone too, not only the brigs, brigantines and snows but their modern successors as well. The Thames is a cleaner today but a much emptier river.

Despite the mounting crescendo of criticism from all directions the only concession to change was an improvement in the discharging methods. In the seventeen eighties the technique of whipping out the coal was adopted. This reduced the team necessary for discharging a collier to nine or ten and employed a technique which became synonomous with the sea-cole trade. Dickens evokes the scene well, . . . "under the figure-head of the John of Sunderland making a speech to the winds, as is done by many Johns' and the *Betsey* of Yarmouth with a firm formality of bosom and knobby eyes starting out of her head . . . . here were colliers by the score and score with the coal-whippers plunging off stages on deck, as counterweights to measures of coal swinging up, which were then rattled over the side into barges . . . . " Four men laboured in the hold and shovelled the cargo into baskets, these were immensely strong and, oddly enough, it was the job of the ship's cook to keep them in good repair. The rest of the whipper's team worked on deck, from a simple, wide flight of steps which was carried as part of the ship's equipment and erected when required. A stout rope led from the black depths of the hold, through a gin-block held aloft by a long spar, and then down to the deck where it divided into four long 'Paddy's bell-ropes', at a point some twenty feet from the deck. These 'bell-ropes' were adjusted to allow for the depth of hold at which the coal was being working by sheep-shanks tied to them. Clutching their individual bell-ropes the team ascended the staging, over-hauling the ropes as they climbed. At a word from the foreman they jumped to the deck in unison and the coal-basket, which had been hooked to the fall of the rope in the hold, whipped up to be caught deftly by the foreman. He disengaged the hook, especially formed to facilitate this and emptied the coals into the vat. Meanwhile, down below the shovellers were filling baskets, digging it out from the black, candle-lit recesses of the hold in an atmosphere of choking

Nineteenth century coal-whippers working on the collier's deck. The cargo is emptied down a wooden spout into the lighter alongside through a gate. *The Graphic*

dust, and dragging them to the cleared space under the main-hatch. The custom was to work the vessel out at the rate of two score (42 chaldron) a day, calculated from the time the meter was appointed. The same predatory 'undertakers' who battened upon the earlier heavers, seized upon the teams of whippers and although a prohibition that undertakers should not also be publicans the law was easily evaded. As a contemporary reporter stated, "In either case the Publicans do not fail to recommend those who expend the largest Share of their Wages at their respective Houses". While the coal-whippers pay of three shilling per 20 chaldrons was princely compared with the wages of a contemporary farm labourer, much was absorbed by the undertakers without whose patronage work would not be forthcoming. There were at this time, it was said, seventy taverns on the north side of the Thames below London bridge, employing coal-whippers and all of the landlords making fortunes out of the earnings of these men. Eventually, in 1843 a Private Bill, taken over by the young Mr. Gladstone at the beginning of his career, established an office where owners and captains might apply for a gang who worked in rotation and finally, in 1851, an injury and pension fund was initiated. At this time nearly all the one and a half million tons of coal discharged in the Port of London was still dealt with by this primitive method; 200 gangs were needed . . . "Some have fustian jackets, some have sailors' jackets, some loose great coats, some Guernsey frocks. Many of them work in strong shirts, which were once white with a blue stripe; loose cotton neckerchiefs are generally worn . . . all have black hair and black whiskers – no matter what the original hue." This method of unloading colliers was not unique to the Pool of London but was to be found all round the coast from Lynn to Shoreham, with minor differences. One advantage of the brig rig was that, once the ship's boat was hoisted vertically up with a handy-billy against the fore-mast, as was the custom, the un-encumbered space over the main hatch could easily accommodate the coal-whippers' gear and work. Recently, forgotten in the corner of a derelict Thames-side warehouse, the complete arrangement was re-discovered. It is hoped, one day, to display it in the new London Museum.

The employment conditions of the humble coal-whippers were not the only reforms to be made at this time as the mounting volume of coal landings forced improved arrangements. In 1803 the City of London had purchased the Coal Exchange in Lower Thames Street and it became a free public coal market through which all coal landed had to pass. Then in order to bring some system to the Pool and its tiers of colliers, masters of the ships were enforced to record their arrival at Gravesend. These were entered in numerical order according to their time of arrival by an Inspector, a scheme which began in 1833. These were sent to the London Office twice a day by river steamer for the attention of the meters for sale on the Exchange. Despite the Geordie's reluctance to lose their habitual right to a near monopoly of the Lower Pool, they were now controlled so that they

Coal drops at Sunderland, early nineteenth century. In the background is the cast iron bridge of 1793, a triumph of the new metallurgy and engineering. *National Coal Board*

moved up the river from one clearly defined section to another. In January 1845 the Harbour Master of the Port of London, Captain Fisher R.N., confidently announced that 900 sail of colliers were ranged between Gravesend and London Bridge; adding that "When they come to Gravesend they are ordered to different stations and as vacancies occur, orders are sent down from London for a number of colliers to fill up those vacancies". I trust that the skippers appreciated the naval precision of the operation. At this time there were trading to the Thames no fewer than 523 brigs owned at Sunderland, 314 at Shields, 252 at Newcastle, while the relatively newly developed port of Stockton had 273 vessels registered there, the majority of them engaged in loading coal cargoes. There were, moreover lesser fleets from Blyth, Whitby and smaller ports whose 'seeking' vessels participated in the trade.

Measurement by the vat of nine bushels was also under criticism. The report of 1830 complained, "Nothing can be so vague as this kind of measurement and it is evident that the Meter must have it in his power, by giving short measure to recommend himself to the Captain". How the meters must have quaked for those chines of pork, lamb and peas and good beef-steak which, by tradition, masters provided for them to help turn the scales. The following year the sale of coal by measure was finally abolished in favour of sale by weight. Masters were not without blemish. Now that most of them were carrying coals for others, and even before this became a general practice, it was widely alleged that they connived at the pilfering of coal which was then openly sold, with a certain appropriateness, at Execution Dock. Further reform of the trade in the eighteen thirties was the repeal of the Customs' duty, the statutory system of meters and meterage and the City's ancient right to levy dues was suspended for seven years. A duty was still payable on coals coming from inland of one shilling and one penny per ton, which reinforced the old system of protecting sea-borne coal. It helped to shield the Northern Alliance and its price-fixing from unwelcome competition. Not only were the Cities of London and Westminster affected but an area within the radius of 20 miles. Established at first to rebuild 52 parish Churches after the Great Fire, it was later devoted to the reconstruction of St Paul's while later the Victoria, Albert and Chelsea Embankments along the Thames also had their cost defrayed in this way. London was not alone in making these domestic levies. Yarmouth collected 'Chapel Duty' of one shilling per chaldron for the erection and endowment of St George's Chapel of Ease. Deal, too made a levy from the coals landed for the building of "A Chapel of Ease to minister to the spirital needs of the Royal Navy". One hopes that the revenue proved sufficient for this important and onerous duty.

The tax to protect sea-borne coal from the impact of an improved system of inland transport seriously affected canal trade. The waterways had tapped the rich

Collier brig *Unity* 1848, by W. Doust 1854. *Science Museum, London*

This battery of drops at which the colliers loaded was built on the Tees at Port Clarence in 1834. They served the collieries of south west Durham to which they were connected by a rail link and were developed by the Clarence Railway Company. In 1839 over a quarter million tons were shipped from the staithes.
National Coal Board

resources of the Midland coal-field and it was intended to make them available to the metropolis by the Grand Junction Canal, launched in 1793. It eventually reached the Thames at Regent's Canal Dock, Limehouse. But the City saw to it that no coal came beyond Grove Park, Hertfordshire, on the canal or by another route, beyond Reading on the Thames. All inland coal reaching London paid a prohibitive duty of about ten shillings per chaldron. This led to a bizarre practice of smuggling coal; a letter from an incensed agent to his northern counter-part acting for Lord Devalal (about 1804) complained bitterly that . . . "from Staffordshire immense Quantities are brought down the canals and smuggled into London, where they can be bought and sold at a much cheaper Rate than Sea Coal, no duty paid on them. They have now got to such a pitch of Hardihood that they publicly advertize them". Officially at least, inland coal only contributed less than one third of one per cent to the fuel which London consumed in the first half of the nineteenth century. As the total of consumption rose, still all arriving by the traditional sea route, discussion became centred upon the question of providing docks for the great collier host, as river moorings were again failing to provide sufficient facilities.

The fleet was a motley one. By the eighteen forties it contained newly launched vessels with the high bulwarks and quarter-boards which distinguished them from their eighteenth century companions, originally built with open rails aft and temporary stanchions amidships. They both favoured the plain 'no-nonsence' stem without the decorative head-rails, traillboards, knee and figure head beloved by less utilitarian craft. It was an austerity they had inherited from the collier cats of the eighteenth century, although the stern was now the square counter of the typical contemporary coaster; the range of dignified windows lighting the master's cabin were now a thing of the past. While the average life of a collier was some thirty years, and a vessel of fifty years was an exception, they numbered in their ranks numerous veterans, finishing their days in a trade where fast passages were of little consequence and a wetted cargo was of little concern. Doyen of these was the *Betsy Cairns* whose story illustrates, if nothing else the extent of human gullibility. Her keel had been laid in the King's Yard, Deptford, in 1690 for William III and launched as the *Princess Mary*. Rigged as a square sailed ketch, as was the style of the period, she served Queen Anne and then George I, although under the king she was confined to providing accommodation for the Court's attendants. Expense was not spared in maintaining the Royal yachts and she was extensively repaired in 1722. In the middle of the eighteenth century she was sold out of the service and consequently her name was changed. Renamed *Betsy Cairns* she traded to the West Indies, owned in London. In 1803 she had further extensive repairs and her topsides were raised to increase her cargo-capacity, the war-time shortage of shipping no doubt making this an economic proposition. She was employed in the transport service from Plymouth, armed with four 12-pounders and a crew of twelve and voyaging to Spain for the Peninsular Campaign. After these exploits she was relegated to the coal trade from the Tyne, owned by a London company and finally passing to George Finch Wilson of South Shields in 1825. Two years later, outward bound towards Hamburg with coals she ran on to the Black Middens rocks, south of the Tyne entrance in a February snow storm. The crew were saved and the cargo salvaged, but this by no means exhausted the potential value of the 137 year old veteran. The press declared her the oldest British vessel and that she was the ship which brought King Billie from Holland; this merely reiterated an old legend rather than fact, for William III had made his historic crossing two years before the vessel had been launched and it was well known that he came in the man-of-war *Brill*. No matter, there was even a superstition that as long as *Betsy Cairns* was safe Protestantism in England was secure! The result was that Orange Lodges proudly claimed fragments of the derelict; crowds hurried to view the wreck and pieces were sold at exorbitant prices all over the English speaking world. A carved figure, part of the vessel's knight-heads, was presented to Trinity House, Newcastle where it still remains and snuff-boxes made from the salvaged timbers given to each member of Newcastle Corporation.

Not all the coal-fleet had such a colourful past nor such an unusual end. But halfway through the century changes, which were to bring about a revolution to the coal-trade and were to sweep away the collier-brig, coal-whippers, the watermen, the heavers, grimy coal-backers and much else, were at hand.

Here comes our waterman and boy,
The constant man we do employ;
His name is Sam, a clever fellow,
But Sam is often a little mellow.

The wind is fair and you must know
Sam did ashore at Greenwich go;
A steady breeze, the river clear,
We went along had nought to fear.

At night we brought up within the Swatch,
Furled sail and set the anchor watch,
Next morning before break o'day
The mate knocked out to heave away.

The wind north-west, and a strongish breeze,
By a reef we did our topsails ease,
The wind continuing still the same,
At night to Yarmouth Roads we came.

And when abreast of Caister Rails,
We brought her up and furled all sails;
We have been here now a week today,
The wind comes fair, we're under way.

Each man a sovereign had in London,
The place that sailors are so fond on;
And he that leads a regular life,
Has good stocks now to please his wife.

And some there are that have got none,
But still it seems the sovereign's gone;
We have passed Dudgeon and seen the Spurn,
Flamboro' Head will come in turn.

Our captain now seems better pleased,
The wind's at South, the pinch is eased
He does expect, if all goes right,
She will be down tomorrow night.

I'm thankful now we're here at last,
At Fairlees' crane we make her fast;
And will be there, I make no doubt
Until the ballast all be out.

From *Description of a Voyage in the Coal Trade,* mid-nineteenth century ballad.

Sunderland harbour from the pier. The piers of many old harbours were equipped with heavy wooden capstans; these were used by the dock-hands to warp out ships in the event of head-winds. *Author*

The economical brigantine rig, shown here in its early form, was adopted towards the end of the days of sail in the coal trade. *From a watercolour by W. Daniell* (1769-1837)

The wreck of the Whitby built *Sir Henry Havelock*, launched in 1859, on Cleethorpe Sands in November 1893, shows the deck layout of the typical nineteenth century collier. The windlass aft of the fo'c'sle companion is particularly interesting. *P. Barton*

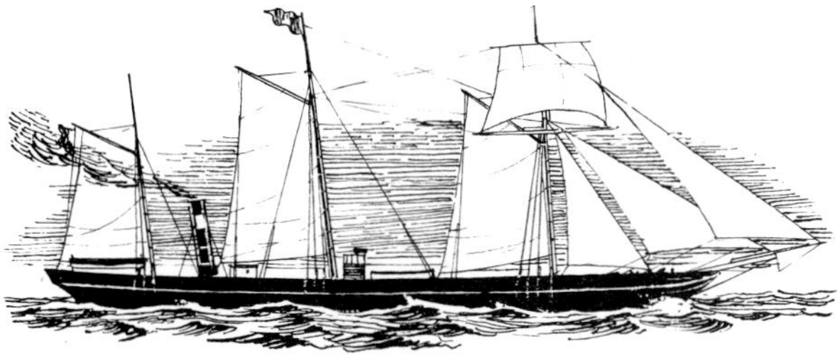

11 The Coming of Steam

"For six centuries the shipowners had had the trade in their own hands and they carried it on, in the present day in the same model as was in vogue at the beginning (hear, hear). What wonder that the coal owners should at last take the matter into their own hands and see if some other kind of vessel would not better answer the purpose of the coal trade? (Hear, hear)."

Report of the speech by George Palmer at the launch of the *John Bowes,* June 1852.

A BOY newly apprenticed to a sailing collier's captain in the mid-forties of the century would by the time he had become a fully fledged able-seaman have seen a revolution in the trade ashore and afloat. Amongst other ancient institutions the very system of apprenticeship for seamen was abandoned in 1849 and the rule of one boy bound to serve for every 100 of the vessel's registered tonnage, was finished. There was much else besides, making a transformation as complete as in any industry and in a century that experienced wide and rapid change. In the north the older mines of the Tyne were losing ground rapidly to expansion in the south, of the Tees ports, served by the Stockton and Darlington line whose completion in 1826 had marked for the whole nation the beginning of the railway age. There were entirely new ports at which the brigs could load. By changing Middlesbrough from a village to a thriving coal-port with the first example of a 'railway town', the effect of steam locomotion was there for all to see. The iron road triumphed again in 1835 when by a joint enterprise of railway and dock construction the port of Hartlepool was developed from a rubbish dump with only three feet of water. Its south-facing harbour and the shelter provided by the Hartlepool peninsular was greatly appreciated by the collier fleet, still relying upon sail. Today its north shore with its octagonal light-house and mature houses, colourful by northern standards, retains some of the character it had in the days of hemp and canvas.

For a short period, between the decline of the Tyne and the rise of the ports expanding with the first stirrings of the steam age, the house-coals mined along the Wear enjoyed great popularity. In 1820 one million tons were shipped away, nearly all in the traditional 250-ton collier-brigs. Sunderland built most of them, at small wood-shipbuilding yards along the banks of the tidal river. They were humble beginnings. Bartram built his first brig, the *Crown* in 1833 for a baker at Monkswearmouth, who held most of the shares. Upon this venture, lasting six months, Bartram and his partner realised a modest profit of £77. Peter Austin, another pioneer along the banks of the Wear, had a repair yard at North Sands in 1826. This was the part of the first wave of shipbuilding expansion associated with Sunderland and it saw the construction of literally hundreds of brigs and snows; the bigger barques were products of the 'fifties and 'sixties. Some, like the *Alderman Pirie* of 1840, the *Enterprize* and *Northumberland,* were built expressly for the coal trade about the North Sea and the Baltic. The *Alexander Wise, Emily* and *Maria Hardy* which voyaged to the Mediterranean and the Black Sea, were only relegated to the coal trade permanently in their old age. The *Triton,* built in 1824, shows how the plain colliers appeared.

The newly founded coal-port which excited perhaps the greatest contemporary comment was Seaham Harbour, between Sunderland to the north and Hartlepool to the south. Its viability was entirely dependent upon the early railway-building of the 'twenties and 'thirties and the arrival in the North East of Charles Stewart, third Marquis of Londonderry. The Marquis had been advanced £35,000 in 1833 by the Treasury for the transformation of an insignificant drying harbour into a coal-port, and he had further resources as an immensely wealthy land and coal owner. Blasted out of the cliff and built complete with piers, lighthouse, houses, shipyard and workshops it was connected by rail with the collieries at Rainton. Within a few years of its opening it was known for 70 vessels to leave on one tide. A century and a half later it still maintains a lively trade and its appearance is evocative of the port when it was first planned.

The imperious Marquis had an important effect upon another aspect of the coal empire of the North East. The economic pressure he could bring to bear and his position of self-sufficiency, his egotism and energy, enabled him to give a final push to the precarious structure of the price maintenance system in the coal industry evolved over the previous century and a half. It had nearly toppled in 1829 when its suspension brought down the price of coal on the London Exchange immediately. Change was in the air and the eighteen forties were a tumultuous decade; new attitudes and profound technical changes were fighting for acceptance. For the first time rapid railway transport was bringing Midland coal to the London market, and this alone was a severe threat to any attempts to maintain

From internal evidence this map of the Durham railways dates from about 1830. The rail link between Rainton and Seaham Harbour is not shown as this was a private link. *Roger Finch*

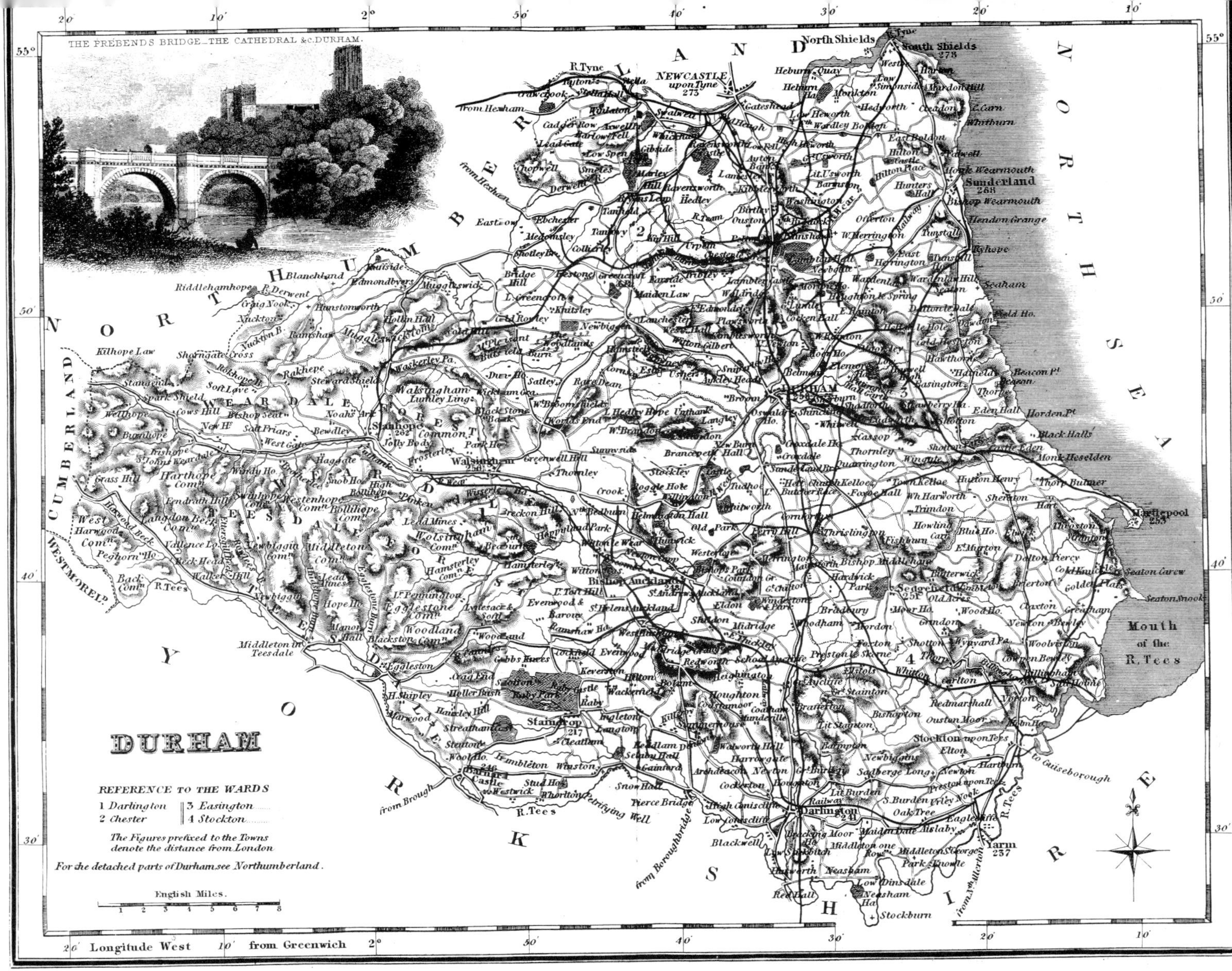
THE PREBENDS BRIDGE_THE CATHEDRAL &c. DURHAM.
DURHAM
REFERENCE TO THE WARDS
1 Darlington
2 Chester
3 Easington
4 Stockton
The Figures prefixed to the Towns denote the distance from London
For the detached parts of Durham see Northumberland.
English Miles.
Longitude West from Greenwich
NORTH SEA
NORTHUMBERLAND
YORKSHIRE
CUMBERLAND
WESTMORELD.
Mouth of the R. Tees
NEWCASTLE upon Tyne 273
North Shields
South Shields 273
Sunderland 268
Hartlepool 253
DURHAM
Darlington 241
Stockton upon Tees
Yarm 237
Stanhope 262
Bishop Auckland
Barnard Castle 246
Staindrop 247
Sedgefield 251
WEARDALE
TEESDALE

prices at an artificially high level. The opening up of the new coal districts in the North East using largely London capital presented another challenge to the old order. Outside the long-established interlocking network of family ties and locally based enterprises, they gave uncompromising competition to the old established pits. The crucial point was reached when Londonderry was expected by his fellow members of the 'Regulation' to subscribe some £8,000 into the communal pool they held. This was used, by tradition, to pay those owners who had fallen short in their production of coal and assisted in maintaining a scarcity situation on the London market. He refused point blank to make his contribution. Such was his economic power and resources that the Regulation of the Vend which had done so much to mould the coal trade, already threatened by rail expansion, the growth of the miners' unions and changes in public opinion, collapsed never to recover.

Railways not only moved coal efficiently from the Durham pits to the new ports; much more seriously they threatened the very existence of the sea-coal trade by making Derbyshire coal available for London consumers. But the railway competition, which started in 1845, did not immediately affect the concourse of shipping in the Pool. In fact in 1850 a collier dock was established at Poplar for discharging coal into railway trucks for distribution by the North London Railway. For a time at least the insatiable appetite of the Londoners for fuel even permitted a modest growth in the tonnage in sea-borne coals entering the port, despite railway competition.

Steam could work to the sailing colliers' advantage. Trinity House now had a steam dredger, working on the worst of the shoals in the Thames, but it nevertheless found the going heavy. The demand was, as ever, for ballast which could be used to stiffen colliers for the return voyage to the North. This was still gathered from the river-bed in a manner which had little to do with the age of steam and iron. A long pole pivoted on the gunnel of the barge had on its end an iron ring supporting a leather bag. This was dragged across the bottom, then winched in by a chain and the operation completed by emptying the contents of the leather pocket into the lighter's hold. The whole archaic process engaged the muscles of a six-man crew of 'ballast-getters'. When the tide made, these lighters would hoist a patched brown lug-sail and work their way back to the Pool, falling alongside the empty colliers on the outside of the tiers. The ballast heavers then loaded the stiffening aboard the brig, first casting it on to a staging set across the lighter's beams and from there on to the deck or through a small ballast port in the vessel's side. To load one-fifth on a vessel's paying tonnage was usually sufficient for safety, but it had to be trimmed in the hold to prevent it shifting, with disastrous results, while at sea. Ensuring that the tonnage the heavers had contracted to deliver was actually thrown aboard was by no means easy. A conscientious mate stood by and timed the process with the vessel's 'half-hour glass', four turns normally measuring an adequate amount, for the heavers worked at a traditionally brisk pace. This would ensure that the ship would, quite literally, stand up to a hard blow. As the difficulty of obtaining ballast increased, the colliers were permitted to take chalk from any point eastward of Purfleet, under proper regulation. This could have a possible economic value for the owners in the North where it might find a purchaser amongst the lime burners at North Shields or, in later years, could be sold for the formation of steamer berths in the tidal rivers. Loam from Erith was occasionally shipped for use in the moulding shops of the Northern foundries.

Having returned to the Northern ports, disposing of the normal ballast caused even more concern for the colliers. Since the time of Henry III at least, the authorities of Tyneside had enforced regulations for the handling of ships' ballast. Having paid Trinity House on the Thames one shilling per ton for it and an additional six pence per ton for putting it aboard, the master of a collier was forced to part with another shilling for every ton of ballast he disposed of on the Tyne. This was paid to the Commissioners; there was a further ten pence for dumping it ashore. Upon reflection the collier-skipper might have found some consolation in the knowledge that the charges had remained unchanged for a century and a half! In an age which was nothing if not ingeniously experimental it is hardly surprising to learn that in the mid-forties of the nineteenth century the idea of using water ballast was explored. Canvas bags were filled with water and lowered into the hold. When safely in port these were discharged into the bilge of the collier and the water pumped over the side; it was laborious, but it was nothing to be compared with unloading solid ballast. In 1851 the coasting brig *Benton* tried the experiment of using rubber bags instead of canvas; they proved more satisfactory, but were inordinately expensive.

The less progressively equipped colliers, upon their arrival at the coal-loading port, tied up at the official ballast quay which was found on all the rivers, clearly marked by its ever-growing hill of dirty gravel. Ballast quays are marked on Hollar's map of the Tyne, engraved in 1654. Another later map dated 1723 shows two, one above 'Jarro Lake' and another higher up at Willington. In 1756, a freeman of Newcastle was licensed to form a ballast quay and charge shipowners 13 pence a ton on all ballast he and they discharged at South Shields. Soon there were nine huge mounds there and 18 from Mill Dam to Jarrow Slake, all producing revenue for Newcastle. They were matched by 150 foot high deposits on the Northern bank and indirectly made an already shallow river worse.

These quays proved inadequate, which is scarcely surprising as by the mid-nineteenth century 900,000 tons of ballast were brought unwillingly into the river annually. A steam winding engine had at least been installed in 1802 at the Willington Ballast Quay. This drew the wagons filled with ballast to the top of the hill and had the distinction of having the young George Stephenson to supervise the machinery. The others relied upon horse-driven winding gins. About 1754 Richard Liddell, connected with shipping and mining interests, devised a new approach. He designed a lighter for disposing of ballast at sea without danger to vessels. Despite the fact that it had the wholehearted support of shipowners and masters and that it would carry ballast out to the 14 fathom line, the scheme was frowned upon. Liddell intended to charge six pence per ton less than the cost of landing it on the ballast wharfs. Newcastle Commissioners, with their traditional myopia, would have none of it. Liddell, nevertheless, has been accorded a certain immortality for his original approach to the problem by the inclusion of the plans of his little sailing lighter in Chapman's magnificent *Architechtura Navalis Mercatoria* published in 1768.

The control of ballast landing for revenue rather than for river maintenance was but one of many restrictions forced by the ancient monopoly of Newcastle upon the other Tyneside communities and interests. Eventually the Tyne Improvement Act of 1851 was carried through in the face of a reluctant Newcastle Corporation. Very soon it had its practical effects and these in turn had repercussions on the

One of the colliers owned at Dover. The heaviness of the rigging indicates it is of hemp. *Author*

collier-fleet. In 1774 it had been said, in despair, that through ignorance, inattention and avarice, the Tyne had been changed from one of the finest rivers of the world to a 'cursed horse-pond'. Even the custom of employing keel-men to clear wrecks and shoals from the river during the slack winter months had been discontinued. Now, at last, a progressive attack upon the problems was to begin.

The shipbuilders of the river rose to the challenges of the new age even if the civic interest was slow to react. Not only did they bring about an extraordinary expansion of shipbuilding, devising new techniques and employing new materials; by their enterprise and invention they saved the Northern coal trade. Iron vessels had been constructed on the Tyne since 1840 when, it is said, the Aberdeen-built *John Garrow* arrived and stimulated the use of the new material for ship-building. An early Tyne ship built of iron was the enigmatically named *Q.E.D.*, launched in 1844 and built of iron by Coutts at Walker as a small sailing ship of 241 tons. She was fitted with auxiliary engines but this did not save her from stranding at the mouth of the Seine and becoming a total loss. The valuable innovation made by her designer was to extend further the idea of water-ballast and Coutts had her built with a double bottom into which water could enter through valves and from which it could be pumped by the ship's engines. If her career had not been cut short she might have been hailed as the first screw collier.

Another vessel which anticipated a variety of developments in ship-building and showed great originality of approach to the design of colliers, was the *Bedlington* of 1841. Only the second iron vessel he had launched, she was built on the yard of Thomas Marshall at South Shields for the Bedlington Coal Company whose pits lay inland and north of the Blyth. Constructed of iron, she loaded 110 tons of coal in wagons in her hold in the manner of a train ferry; an idea which may have been developed from the early experiments of loading wagons on board keels in use on the Wear. The difficulties of transporting coal to the waiting ships further north at Shields would, it was thought, be overcome by loading wagons and then discharging them with a steam derrick carried aboard. Water ballast-tanks were fitted for use on the return voyage along the coast and up-river to the Company's wharf on the north bank of the winding Blyth. Hardly surprisingly when one considers the variety of radical ideas involved in the *Bedlington*, the venture was only partly successful. The new and partly understood technology of building vessels in iron and the problems to be solved when steam propulsion was installed in ships, resulted in continual and expensive repairs. Navigation on the Blyth was hazardous with its sandbanks, shoals and difficult entrance. Although strengthened with timber and extra plates, the hull was continually giving trouble from longitudinal strain and finally the *Bedlington* broke her back on a sand bank in the Blyth. The designers had made things doubly difficult for themselves by placing the boilers in the bow and the machinery aft. With a shallow hull already dangerously long for its beam, the long, open hatch must have led to further problems. The keel of the *Bedlington* measured 134 feet, its beam only 25 feet and its depth of hold was 8½ feet; hogging and the rupture of the steam feed pipes was inevitable. One can only admire the tenacity of her owners in running her for nearly a decade and the good fortune that followed her on her coastwise voyages between the Blyth and the Tyne. The *Bedlington* did not end her days on the Blyth sand-bank, indeed it proved to be only the opening of an adventurous career. She was auctioned and when salvaged towed away to be extensively rebuilt and employed on the river Forth as a ferry and then in 1853 voyaged out to the Crimea to be sunk by Russian batteries the next year. There was a vast requisitioning and hiring of merchant vessels to act as support vessels in the Crimean War, ranging from clipper ships to tiny sloops, but few of them could have had as interesting a career as the *Bedlington.*

The transformation of collier-master from shipping coal and combining the role of management, professional seamen and entrepreneur to become a freighter of others cargoes had gone on apace. In 1850, a Newcastle colliery owner Charles M. Palmer, reviewing the deteriorating situation for sea-coal on the London market, in the face of railway competition, decided he must radically change the shipping facilities available to North East coal producers. The menace of competition from the Midland and South Yorkshire coal reaching London hearths and furnaces by rail was growing; now there was even discussion of a Newcastle and London Coal Railway. He must have known of the *Q.E.D.* and another pioneer built at South Shields in 1847, the iron *Conside.* Palmer, not a dilatory man, in conjunction with his brother formed a shipbuilding company at Jarrow, taking over an old yard which had been building wooden ships for fifty years. The enterprise was styled the General Iron Screw Collier Company and was launched with forecasts of handsome dividends, not an unusual promise to the mid-Victorian decades. But in this case the assurances were to be amply fulfilled. Two months after the keel was laid, in June 1852, the first successful sea-going, iron-built screw propelled collier was launched from the yard of Messrs. Palmer, Brothers and Co. In the deathless prose of the Victorian journalist "as the vessel was majestically marching to her desired element, Mrs. C. M. Palmer dashed upon her sides a flood of blood-red wine and christened her, in honour of the High Sheriff of the County the *John Bowes*".

The *John Bowes* had the double-bottom which enabled her to use water-ballast, pumped out by her main engines before loading a cargo; a device that had been pioneered by the ill-fated *Q.E.D.* She had besides this a 60 foot long hatch to facilitate loading; by moving the vessel fore and aft under the drops on the coal staithe no expensive and time-consuming trimming of the cargo was needed. The sailing colliers did their best to avoid this backbreaking labour by placing a sheet of iron under the cataract of coal and so dispersing its flow as widely as possible, but it had only a limited effect. The 150 foot long hull of the *John Bowes* was fitted with engines built at Newcastle. When originally launched she was rigged as a three-masted topsail schooner and the hull terminated in a fine clipper bow, jibboom and bowsprit and there was very little about her that we associate with the word 'collier'. But so much innovation in design and her capacity to make the voyage to London in an unheard of forty-eight hours, would have been pointless if she had, upon her arrival, been forced to rely upon the reluctant gangs of coal-whippers, restricted by tradition to their 80 tons per day. Here again, change had come. Another Tyneside pioneer William Armstrong had erected six hydraulic cranes on the quay of the West India collier dock discharging coal into waiting railway trucks and these were capable of delivering upwards of 2,000 tons daily. But this equipment was expensive and could only prove an economical investment with constant employment from a steady stream of colliers providing a regularity of arrivals from the North. Such regularity was notoriously impossible as long as the trade depended upon sail; the situation is well illustrated by the details of a great armada of merchant ships which entered the Thames in the same month as the *John Bowes* made her maiden voyage. Delayed by headwinds and calms, a fleet numbering 346 sail came up river in the space of two days, of whom 124 were colliers. It was this situation of embarrassing numbers followed by an equally crippling famine that the *John Bowes* and her successors could solve. The newly launched collier took forty-eight hours to reach the Thames-side docks, discharged her cargo in eighteen hours and in two days was again on the Tyne under the coal-drops. Loading another 650 tons of coal was accomplished in only four hours.

The *John Bowes* cost her builders £10,000; a trio of wooden brigs, newly built and fitted out ready for sea could have been launched for the same price. But the steamer had achieved in five days the accomplishment of three brigs working for a month. Moreover the coal arrived in a less broken and therefore more saleable condition and "steam-boat regularity" became a phrase indicating the highest approval of efficiency. The *John Bowes* was rapidly followed by others; two were in frame on the shipyard at the time of her launch. The *James Dixon* was capable of loading 1,200 tons in four hours. She performed the work of sixteen sailing colliers, with crews totalling 144 men; the *James Dixon* had a complement of 21 although admittedly this included highly paid engineers. Within a few years there were eight Tyne yards capable of building iron ships and by 1854 36 steam iron colliers were employed in the London coal trade. Stockton launched her first screw steamer in 1854, the *Advance* of 107 tons, to be owned there and later at Middlesbrough, but their first screw collier was probably the *George Stephenson.* The famous firm of T. & W. Smith launched the iron screw *Chasseur* in 1853, intended for the coal-trade only to have her bought up by the Government, as was the *George Stephenson* just as colliers had been for centuries, for service with the Royal Navy. She went out to the Crimea to become a floating workshop and to attend the demands of the fleet. Despite the gloomy forebodings that iron ships

The launch of the screw-collier *John Bowes* at Jarrow, July 1852. *The Illustrated London News*

would corrode, cause incurable magnetic distortions and the bottoms of the vessels would foul, the North East led the rest of England in this technological revolution. But iron colliers were not immune from storm and sea; they too had to anchor occasionally in the lee of the Scroby sands in winter gales just as the sailing-colliers did, and delayed by strikes or mechanical failure they represented a vast amount of non-earning capital. But their combined effect at once saved and revolutionised the sea-coal trade.

The *John Bowes* had been iron-built to last. She was given new, more powerful engines in 1864 and again in 1883, for the heavy gauge of the plating used on her hull and the capacity of iron to withstand corrosion, made this expense worthwhile. On the second occasion of the replacement of her engines she was fitted with compound engines of 66 nominal h.p., together with a new boiler, at Dumbarton. By this time she was relatively small and therefore uneconomic in comparison with the standard colliers used in the North East trade, and she was disposed of in 1898 to Norwegian owners, who changed her name to *Spec.* She was sold again two years later to the Swedes who renamed her *Transit.* As the *Transit* she returned to Palmer's yard at Jarrow in 1905 for repairs. Now shorn of her elegant clipper bow and schooner rig she set only a few rags of smoke-stained steadying-canvas. Three years later the old collier changed hands for £2,000, this time registered at a Spanish port, Gijon, as a general cargo vessel and at one time used for cable-laying between Dover and Ostend. In November 1933, then named *Villa Segas* she foundered during a gale; in the same year Palmer's shipyard at Jarrow, victim of the Depression, sank bankrupted, beneath the wave of shipbuilding 'rationalisation'.

The appearance of the steam collier with such unqualified success had far-reaching effects on the Thames as well as on the Tyne. Because of their rapid 'turn-round' they remorselessly reduced the number of sailing vessels employed in the coal-trade and consequently the congestion in the River which, however picturesque, was clogging the commerce of the Port. Machines for unloading ships had been devised as early as 1757 but were never able to break through the strongly entrenched 'undertakers' who provided manual labour to do the work. In 1805 a machine had been invented by the Cornish engineer Richard Trevithick. It was banned by the coal whippers, who threatened to go one further than the Luddites and drown the inventor as well. We can have a certain assurance in the practicality of the lost design of Trevithick, but perhaps the usefulness of another advertised in 1830 may be viewed with a certain scepticism. Driven by steam it delivered 20 chaldrons per hour, measuring and weighing the coal at the same time. Taking up no more space than a half-ton cask, it could unload and load, lift ballast and the inventor concludes by way of an afterthought that "it might also pump the ship in a gale of wind". Despite the availability of such ingenuity the tradition of unloading by gangs of whippers continued as late as the eighteen sixties aboard all steam colliers discharging in the Pool, although more than one gang worked at a time. As late as 1856 one and a half million tons every year were discharged by hand in the Pool, at 6½ pence per ton, and as late as 1863 three-quarters of the sea-borne coal was still carried by sail although ten years later the victory of steam was almost complete.

Such was the capital investment in the screw-colliers that coal-owners naturally gave them priority of loading at the coal-drops, leaving the sailing-craft at a further disadvantage. A wait of a few days meant nothing to the brigs when the journey to the Thames might be prolonged to as long as a month, it was argued. On the Thames the steamers led to the abandonment of the 'turn' system with its registration and reporting at Gravesend. Instead, a signal station, which recorded the passing of colliers was built at Tilbury in 1867 and the Gravesend office merged with it, much to the disadvantage of river-side pubs. Another change came when William Cory decided to avoid the congestion and delay of the upper Thames by purchasing an unused floating derrick which had been designed for salvage work and using it as a floating coal depot. It was moored in Gravesend Reach where colliers could be discharged alongside by a battery of steam cranes. Lighters were then loaded as they were required and towed to destinations along the River. It finally broke the strangle-hold of traditional discharging methods in the Pool and by 1885 manual unloading was exceptional.

The Corys led the campaign in the eighteen thirties to reduce the cost of coal handling and to lessen the statutory charges on coal cargoes. They were among the first to charter the new steam colliers and the adaption of the Thames Ironwork's unsuccessful derrick in 1861-2, which William Cory named *Atlas* was to ensure the profitability of the vessels he had chartered for the coal trade to the Thames. Beside the hydraulic cranes and the apparatus for screening coals the giant float had a gas generator for lighting operations as they proceeded round the clock. The labourers on *Atlas* were worthy successors to the traditions established by the gangs of coal-whippers and coal-backers of the Pool. They were taken off to the float by tugs, each man with his gallon demijohn of beer and ample supplies of bread and cheese and these refreshments were reinforced by a regular waterman known as a 'Coolipper.' He ferried supplies aboard during working hours. With the help of William Armstrong's cranes the men cleared 60 tons per hour, compared with the coal-whippers average of 25 tons, so I have no doubt they needed it.

The *Atlas* became the *Atlas I* in 1875, when the *Atlas II* was built on the Tyne for similar work. By this time there were screw-colliers carrying over one thousand tons of coal and their numbers were growing rapidly. The General Iron Screw Collier Co. had launched 11 steam vessels between 1853-5 alone. Each one could make between 30 and 36 voyages in one year compared with the sailing colliers modest 10 or 12. Another aspect of change, and one equally important to shipping and its organisation, was underlined when, in 1851, the Coal Factors on the Exchange asked that their instructions from the Northern coal-owners should be despatched by the new 'electric-telegraph.'

It is difficult to draw up a balance sheet of profit and loss, of improvement and deterioration in the lot of a collier-seaman as the change to steam quickened. Perhaps a primary consideration should be given to the fact that steamers were infinitely safer. Nevertheless the grind of a continual and unremitting round; loading of the staithes, clearing for sea, the journey to the Thames, discharging and then, with the hull pounding and the screw thrashing, without a pause, the passage back to the Tyne, produced a treadwheel of existence. A screw-collier arriving at Tyne Dock at one or two o'clock on Saturday afternoon was expected to, and did, "save the week-end" and load and sail the same night. Overtime was unknown and working hours were whatever the master thought necessary, as on the sailing-vessels. But the risk of ship-wreck and death by drowning or exposure was certainly less.

The British Merchant Shipping Act of 1850 stipulated the minimum living space each seaman should be entitled to for his quarters. It was determined at approximately one eighth of the amount allocated by the contemporary War Office as the minimum for a soldier's barrack accommodation! The Act, although a poor tool with which to shape a great and expanding industry, did at least introduce a system of certification by Board of Trade examination for those seeking to command a British vessel, permitting men who had proved themselves competent by experience to be issued with qualifying documents which became known as "Certificates of Servitude". There must have been amongst the latter men who, as boy apprentices, had fought privateers, without any qualifying papers, during the war against Napoleon. At least one successful captain of a screw-collier of the nineties could neither read nor write. But then, as his colleagues said, "he could count pound notes alright".

The Act was obviously a very necessary piece of legislation, but it made it more difficult for an enthusiastic boy, lacking, as many in the nineteenth century lacked, a basic education from reaching a position of responsibility. Whatever other social divisions existed in life in the nineteenth century the dividing line of mid-Victorian society was the qualification of respectability. The steam-collier crew, in their crowded foc's'le, which all too easily degenerated into a floating iron slum, were certainly placed beneath this demarcation line. The rough and ready egalitarianism of the sailing coasters and short-sea traders was passing. Aboard the brigs dangers were a shared experience with, if one was lucky, grog all round when they were fought through; food was prepared in common and the margin between the wage of an able seaman and a mate acceptable to both. Now no longer was the captain's son or nephew to be found in the foc's'le or a casual onlooker unable to distinguish between able seaman and mate, while a feeling of group identity was diminished.

The change was accentuated by a tendency towards the owning of vessels by joint stock companies, even before the transference of the coal-trade to steam. On the Tees the Hartlepool Shipping Company, founded in 1835, owned some 18 vessels while on the Tyne, at South Shields the Young families fleet, with its distinguishing 'house flag' of an iron frying-pan wind-vane on the fore-mast, had assembled a considerable fleet by the fifties. This was a change from a managing-owner form of control, when an individual, charging expenses and commission, conducted the main business of the ship which he owned in conjunction with a small group of individuals. Vessels had now become too expensive to raise the initial capital required in this manner. With big companies the relationship between the employed and employer became remote. When the skipper was a considerable share-holder in the vessel and was his own manager it might lead to a careful consideration of the expense of the salt-beef, but a deck that needed re-caulking leaked into everyone's bunk. Nevertheless aboard a screw-collier there was warmth and shelter unknown on the little brigs, no topsails to be reefed in a gale, no tiny vulnerable caboose on deck as the only resort of the cook. Even if scalding tea at all hours produced stomach ulcers and sweating iron foc's'le bulk-heads led to a high incidence of tuberculosis, there was no frost bite or gangrene.

By the second half of the century the shipping industry was complaining of the deterioration of the quality of seamen in both sail and steam and blaming it upon the rapid decay of the sailing-collier. While some of it was exaggerated there was an element of truth. No longer were there men in the foc's'le who had served an apprenticeship, looking forward to the time when they would become skilled masters of their craft and who made up the backbone of a collier's complement. They now tended to be young men who wished to escape, for at least some time, from the work disciplines of a remorseless industrialism ashore. Some with a background of sea-life succeeded. Walter Runciman whose book *Collier Brigs and their Sailors* tells of this period with a wealth of anecdote from first hand experience, rose to become a commander of the barque *F. E. Althouse* at the age of twenty-one and later a wealthy ship-owner. Others became pilots, achieved command of new steam-colliers or, with their skills and resources sharpened under sail were rated bosuns on ocean trading vessels. But many drifted on and away with little regret.

Breaming a collier. *Drawing by the Author*

The *Herbert*, typical of the Whitstable Shipping Company's collier barquentines. Roger Finch

12 The Last Days of Sail

THE miscellaneous collection of craft which freighted the last coal-cargoes along the coast from the North East under sail carried every sort of rig and were drawn from a wide variety of sources. Before we turn to them and how they survived we might take a last look at the traditional collier-brig or snow before they finally bowed to the competition of steam on the Thames.

The hull was deep and its greatest beam was forward under the cat-heads. There was little pleasant sheer to the rail, but a purposeful solidarity about everything; nevertheless the stern and square counter were usually contrived to give a relatively graceful termination to the hull. The tonnage rules of the first half of the century tended to influence builders towards a relatively narrow, deep hull and this affected, somewhat adversely, both the looks and sailing qualities of the brigs. The bulwarks were high and capped with quarter boards aft; there were wide hinged ports in the rails which acted as freeing gates when solid water came aboard and through which were passed the big warps and springs when the vessel was moored. Ranged in tiers, or against tidal docksides, the wear upon topsides could be murderous; expendable fenders of brushwood liberally coated with 'slush' from the cook's galley were used to reduce the damage. Unlike their deep-sea contemporaries the shrouds were rarely set up from outside channels but at topgallant rail level. On deck there was always a simple winch for general work and to which purchases were led, usually at the foot of the foremast. The steering wheel aft was unprotected by any wheelhouse and there were collier-brigs about the coast until after the end of the nineteenth century that were tiller-steered. The little *Sarah* of Shoreham kept hers until dismantled at Southampton in 1910.

The colliers were always renowned for their smart anchorwork and the windlass was a most important part of their equipment. So much depended upon it that it retained the simplicity of its eighteenth century counter-part; there was then an absence of mechanical complication ensuring a reliability in an emergency even if there was little reduction of physical labour. However, by the eighteen sixties and seventies the more advanced pump-handle type was fitted. Known as Armstrongs Patent it certainly shortened the cable far more quickly than its predecessor but could not equal the old, simple, barrel windlass's ability to break-out an anchor.

The boats, often two, were carried on the main hatch. Thirty years earlier a pair of wooden davits over the stern had been almost universal and provided what must have been a more convenient method of shipping the smaller of the two. The increasing congestion of the London river and dock regulations, which discouraged it, produced the change.

Change aloft in the pattern of rigging had been if anything less remarkable. Since the eighteen hundreds the colliers had at least acquired bulwarks and rails; in the sails and spars little had been gained. The introduction of wrought iron trusses to support the yards marked one of the few easily noticeable improvements and these had replaced the older arrangement of parrels and rope ties. More important it enabled the yards to be braced up sharply when sailing close-hauled. Another improvement was the use of chain where the running rigging suffered most wear. Since the eighteen twenties this had been gradually introduced for topsail sheets, where the failure of rope could bring disaster. The introduction of double topsails in order to make the reduction of sail easier with smaller crews became common in the seventies. The old *Remembrance* was so fitted at the time of her loss, but most of the brigs were gone before they became common and they depended upon the traditional three rows of reef-points on a deep single topsail. The *Commerce* of Newhaven was the last brig on the coast to carry them and she was hulked in 1906. The introduction of steel wire standing rigging also appeared too late for the vast majority of the colliers. It was fitted to superior newly-built vessels as something of a novelty in the late eighteen fifties. However as the traditional hemp gear wore out and required replacement many of the remaining coasters and Baltic sea traders of the North East were re-fitted with flexible wire rigging. The considerable difference in the diameter of the two materials is always apparent in old photographs of shipping. By maintaining more rigidity in the masts it improved the sailing qualities of any vessel, especially on the wind. A trysail of extra stout canvas was universally carried aft of the fore mast, to be used in gale force conditions when all other sails were stowed; and despite their association with lordlier vessels stunsails were still in use. No doubt the bulky hulls of the colliers needed all the lugging along they could contrive. Until the last days of the sailing-colliers these flying kites were carried in fair weather, at least on the weather yard-arms of the foremast when the wind was aft. Except for the brigs reduced to coal carrying as a mere expedient between more lucrative voyages to the Black Sea or West Indies, a royal was only carried on the main mast. Indeed for the brigs permanently condemned to coal-drudging it was not uncommon for the fore topgallant mast to be sent down for the duration of the winter months.

To balance the sail-plan it was necessary for the brigs to carry enormously long bowsprits and jibbooms, fully one half of the length of the hull, another mark of their traditional conservatism which they maintained until the end. To conform with port regulations they brought the jibboom inboard when they were upriver of Greenwich. The little collier *Bee* survived to be photographed carrying her jib on a traveller, a nicety of rigging practice that belongs more properly to the eighteenth century.

The *Bee,* with her antique rig, worked her way backwards and forwards loading flour for the North East and then returning to that aquatic corner of Essex known as the Walton Backwaters, a commercial exchange which had beginnings in the sixteenth century. By this time although the Thames rarely found a place in its economy and market for the hundred and fifty or so tons freighted by such vessels as the *Bee* there were still harbours and havens remote from industrialisation that did.

The tradition of black painted ports along the topsides of the colliers was continued by many of them until the very last. It originally dated from the days when it assisted in the deception of privateers, giving at a distance, the illusion of an armed vessel with gun-ports. The gasworks now provided cheap and ample supplies of tar and 'black-varnish' to maintain the tradition. An alternative was a broad white band; both were changed to blue as a sign of mourning upon the death of an owner, and remained this colour until the vessel was refitted. The quarter boards aft were sometimes scrapped and varnished, but invariably carried a name-board suitably carved and embellished. Inboard the bulwarks were treated with brighter colours despite their constant disfigurement with every sort of industrial grime; green stanchions setting off white boards or pale blue and red were common. Decks were painted or tarred and sanded to give a good footing. Along the deck was ranged the chain cable, secured by lashings to ring-bolts and ready for immediate use in an emergency with enough forward of the windlass to enable at least one anchor to be let go. When riding out a gale at anchor in the Downs or in Yarmouth Roads it was essential that some of the weight was taken off the long suffering windlass bitts. This was effected by shackling on a thick grass spring to the cable close to the hawse-hole and leading it aft to one of the two massive sampson posts, normally used for belaying ropes in harbour. This was heaved taut by tackles shortened on the cargo winch and then, with the chain cable slackened off, some of the strain was absorbed as the brig's bows rose and fell to the seas.

The *Ida* of Ipswich, 136 tons reg. Typical of the ketch-barge built during the second half of the nineteenth century for trade along the East coast. Many like the *Ida* carried coal from the North East coal ports to the smaller harbours and riverside 'hards'. *National Maritime Museum, London*

The oak 'harness cask' was a feature of the deck arrangement aft, normally locked and in a position where it could be supervised by the captain. It was usual for a butcher to supply a half-side of beef; having shares in the vessel this he might well consider his privilege. The beef was dismembered, liberally rubbed with coarse salt, then placed in the brass bound cask and well covered with a solution of brine. The constant movement of the vessel at sea was said to ensure a good standard of preservation and savour. Such was its reputation that it was not unknown for landsmen to purchase small quantities of salt-beef from coasters as they lay alongside the quay! A pound of meat per day was the accepted amount allowed for each man in the crew. Newcastle crews traditionally provided their own sugar, tea and butter; elsewhere these were provided by the ship and the pay adjusted accordingly.

A list of freight-rates published by a firm of Newcastle shippers in 1830 includes places which were in effect open beaches among numerous little ports which were rarely frequented by any vessel larger than a brig. Included on the list are Hastings and Eastbourne, with no form of protection for a collier discharging on the open beach and not surprisingly offering some of the highest remuneration to compensate for the considerable risk of total loss to both cargo and vessel.

The shorter voyage to Southwold and Wisbech apparently produced a much better profit than a freight for the Medway — until one recalls the difficulty of entering the Suffolk port's harbour and the expense of towage and the possibility of delay on the Nene. As well as the open beach work already mentioned, the freights paid for landing at Deal, where beach-discharging lasted until the eighteen seventies when the red-brick coal store adjacent to the shore was demolished and replaced by the Royal Marines swimming-bath, Worthing and even Sidmouth in Devon, are quoted and because of the inherent danger, were awarded commensurately higher rates. Brighton, where the colliers lying off the shelving shingle beach was brilliantly recorded by John Constable, secured the highest freight of all, 15 shillings per chaldron. There the brigs discharged in a primitive haven, close by the present Pool Valley depression and later on the open beach, between Ship Street and West Street. The generous reward for undertaking such work was swallowed up by the high level of insurance which became so exorbitant that it brought an end to the trade in the eighteen thirties. Bridport is also mentioned where the strange anomaly of a turnpike gate between the station and the town inhibited the sale of land-borne coal after the railway arrived in 1857 and enabled sailing-colliers to work there profitably. The westernmost port mentioned to which it was expected to send North East coast coals was Falmouth. By the eighties all these rates had fallen some by as much as half, the railway having reached even the Western peninsular.

These modest markets for sea-coals and the larger ones of Yarmouth, Dover, Shoreham and Portsmouth survived into the post-railway age, depending for their existence upon a variety of precariously maintained historical accidents. An important example of this was the manner in which gas producing plant had been sited decades before the railways had spread across Eastern and Southern England. Those that were built in the ports had quite naturally been constructed where they could receive sea-borne coals. This made it worthwhile for them to continue to contract for fuel to be brought by water long after the railway had arrived. To a lesser extent this also applied to other long-established quayside industries such as malting. There were also harbours, like Lyne and Bridport, small and shallow, which did not warrant improvement. The steamers did not deign to use them because of shallow entrances and primitive unloading arrangements and this left a gap which the sailing-vessel could fill. By cutting their expenses to a minimum and purchasing second-hand tonnage the sailing collier owners could, for a while, survive in their traditional trade.

One of the most successful fleets which managed to survive into the twentieth century in this way, was the one assembled by the Whitstable Shipping Company. Its survival was ironic for, while the railway had proved the beginning of the end for so many sailing-colliers, Whitstable harbour had been constructed in 1832 as the first in Southern England to be built with a rail link. Yet it developed to be the home-port for a fleet of some one hundred ships, most of them spending their days tramping chalk-ballast to the North and pumping coal back. At Whitstable coal was discharged as late as the nineteen twenties by the traditional jump worked by coal-whippers into coal wagons to be distributed all over Kent. The last fleet of sailing-colliers, their tall, square-rigged masts were to be found rising above the ketches spindly top-masts and barges' sprits among the grey industrial backs of any small port between Sheerness and Exeter. Whitstable ships and those from nearby Faversham, which was the port of registration, continued to bring coal to all the South and South Eastern ports until 1916 when war-time conditions laid-up those remaining ships which had not become war casualties.

The company had its own slipways and bought vessels which it re-fitted and re-rigged. In this way a heterogeneous collection of ships was built up which, taken together with the other Whitstable colliers outside the fleet, contained representatives of every sort of small trading vessels which once graced our shores. They owned barquentines like the *Herbert,* built in Norway, and the *Zebrina,* one of the last big barges which at one time had sailed across the Atlantic and worked on the River Plate. The brigantine *Joseph* had been built on Prince Edward Island, Canada in 1877. Another Faversham owned collier, the brigantine *Raven,* one of the last to survive, was also Canadian built, as was the bigger *Ibis,* which had been re-rigged as a barquentine. This change reduced the size of the

Two of the Whitstable Shipping Company vessels refitting at their home port c. 1900. *Author*

individual sails, particularly the fore and aft mainsail and enabled a crew of six to handle them. There had been many brigantines in the coal-fleet towards the end, both in the North East as well as the South East, the majority built at Prince Edward Island. They made their appearance on the North East Coast as steam came in during the eighteen sixties, their competitive price made possible by the vast resources of timber and the lower labour costs which had made them an attractive proposition to buyers looking for replacements for the ageing brigs. Their brigantine rig enabled a crew to be reduced to four men and a boy. They sailed in company with the older brigs with gas coals and naval bunker coal to the Baltic, where the steamers had not yet penetrated and the quantities required in one shipment were small. The trade to the French ports, Honfleur and Rouen, were still available for, as in the older British harbours, the primitive unloading equipment would delay a steamer too long to make the voyage a profitable one. Sail could always afford to wait and make the best of it by painting, tarring and over-hauling the rigging.

Besides these the Whitstable fleet contained boomsail barges, ketch rigged and loading between one and two hundred tons. They had been evolved to meet the requirements of a cheaply built coaster, with a shallow draught and low running costs; the first were launched in the eighteen fifties. They multiplied and had almost entirely eliminated the smaller schooners by the end of the century. They were coasters in the general trade loading anything from tree-trunks to fresh herrings, but many worked regularly for years in delivering gas and house coals all round the East and South coasts and all carried black-diamonds at some time or other. Like all the sailing vessels that traded the North East coal ports they were forced to wait until the steamers were loaded before they came under the shutes. This might mean that a sailing vessel could wait, idle for a week or more until bad weather or fog, or perhaps a mechanical failure, held up the steamers. Seaham Harbour was rather less busy than the larger ports and when an easterly gale blew, booms were placed across the entrance to prevent a surge sweeping into the harbour and wreaking havoc amongst the tiers of craft. Seaham frequently features in the pages of the cargo books of these last of the sailing colliers and it was here that a primitive coal drop remained to be finally dismantled and rescued for re-erection at Northern Regional Open Air Museum at Beamish.

The little quays along the Solent and Chichester harbour saw a motley fleet of ketches, schooners and brigantines bringing coal to Itchenor and Dell Quay for the local community. Typical of them was the brigantine *Emily Anne* built at Southwick, Co. Durham in 1867 and sold south when the change to steam began in earnest. The firm of Porstmouth coal merchants Henry and James Ash owned sailing colliers, amongst them the brig *Waterwitch,* launched in 1872. She earned a great reputation in the trade from the Northern coal ports and in 1884, in order to economise in crew and gear, she was converted, like many of her contempor-

Until the end of the nineteenth century Whitby, with its 'Haven under the Hill' was a home for wooden ships. The brigantine *Emily Anne* of Poole, built at Southwick, Durham 1867, was owned by T. Smart of Bosham, Sussex when this photograph was taken. In the background is the barque *Sylvan*, launched at Hobkirk in 1857. *Sutcliffe Gallery*

aries, to a barquentine. She served in the coal-trade for over thirty years and earned a high reputation before being sold to owners at Fowey for general coasting work. When, after many adventures, she was bought by Baltic owners in 1939, she had the distinction of being the last British square-rigged cargo carrying sailing ship.

The re-rigging of the *Waterwitch* was an example of the efforts that were forced upon owners and masters of coastal sailing vessels to achieve operating economies which could enable them to survive. Brigs were converted to brigantines, rigged only with square sails on the fore mast, or like the *Waterwitch,* to barquentines or even topsail schooners. They served areas which were largely unindustrialised and where tempo of enterprise was still relatively slow. Cheap, unorganised labour was available for maintenance, unloading and manning and this helped to perpetuate an ancient trade where elsewhere it was an anachronism. The crews, it was said of the last sailing-colliers, "went to sea for the grub and earned money working out the cargo." Many of the ports of the South East and East coasts had a long tradition of wooden shipbuilding which had been driven to the wall by the iron and steel construction of the Tyne, Tees and Wear. There were still the ships; the managerial skills and maintenance facilities, sail lofts and slipways also survived. As deep-water trading under sail became less and less profitable the ships might still be found a worthwhile employment in the coal trade. There the investment in good workmanship and fine materials might earn a modest dividend until the hard life eventually drove them to the breakers' yard.

Besides this re-deployment of displaced tonnage there was also a response by the new building of sailing vessels for the coal-trade. These were ships which represented one of the few radical departures from traditional wooden ship-building in Britain, an industry in which technical progress was neither rapid nor revolutionary. A new and easier path to profit was sought by building cheap hulls which would serve well enough for bulk cargoes. These were greatly enlarged versions of the hard-chined Thames barges and their appearance in the coal trade date from the first years of the serious steam competition. It was an attempt to produce a vessel of moderate draught, capable of sailing without the expense and trouble of ballast and cheap to produce by reason of the flat-bottom. Throughout the nineteenth century 'compass-timber' for the frames of wooden ships was becoming increasingly expensive; by reducing this to a minimum it enabled drastic economies to be made in the expenditure on skilled workmanship and materials. The earliest of these big barge-colliers I can trace is the *Leading Star,* launched in 1856 at Colchester and, despite the plebian form of her hull she was rigged as a barque of 210 tons. Her maiden voyage was to Sunderland where she loaded coals for Ipswich. The last to be built was the *Enterprize* of Yarmouth, launched there in 1891, of 109 feet and 190 tons gross. She survived until the First World War, rigged as a three masted topsail schooner, working in the coal trade to her home port, delivering bunker coal for the fishing fleet.

In the years between the launching of these two vessels collier barges of up to 800 tons, fitted out with every description of rig, were built. Many were registered at Rochester and the *George Smeed* and *Eliza Smeed* were two early examples owned there. These and others of the same type were built at Sittingbourne, Kent, by George Smeed; the largest was the *Esther Smeed,* built in 1868 and said to be capable of carrying 800 tons. It is difficult to believe that vessels as large as this could carry the barge's traditional lee-boards but certainly the three masted schooner *Friendship* loading some 420 tons had them. She was one of the fleet of colliers owned by Crundall working from Dover. Vessels of a similar type were built at nearly every one of the traditional barge-building centres, Whitstable, Harwich and as far south as Emsworth where Foster launched the wooden barquentine *Thorney Island* in 1870, loading 350 tons.

The longest-lived of these extraordinary craft was the *Emily Smeed,* which could proceed to sea at neap tides over river bars and carried her five hundred tons on a draught of a little over thirteen feet. This was less than the typical south-coast brig only capable of stowing away little more than half that amount. The *Emily Smeed* was rigged as a barquentine with a patent-roller-reefing top sail and was owned at Newhaven by John Bull, bringing gas-coals up from the north for many years, until she was sold to become a lighter at Aberdeen in 1903, lasting much longer than the majority of her kind.

Newhaven, eight miles west of Beachy Head, is the best tidal harbour between Portsmouth and Ramsgate. It was one of the first stations for the packet steamships and sailing-colliers brought bunker-coal for them besides house coals from the Tyne and the gas coals from Hartlepool. Newhaven's own fleet of colliers was the Bull Line, whose house flag of a bull's head on a white ground helped to engender a touch of self-esteem in a trade where it was often sadly lacking. The group was largely composed of old deep-water vessels and amongst them was the snow *Pennine* built at Sunderland in 1867, lost off Yarmouth in 1897 in the coal work and the veteran *Commerce*.

Further to the West, Shoreham's fleet of wood and iron vessels trading all over the globe and dwindled away by the end of the nineteenth century to a varied collection of coasters, generally finding employment in the coal trade. They had the advantage of the one and a half mile Southwick canal, constructed in 1855, which brought the brigs and schooners to the back door of Brighton and its expanding market for both gas and house coals. Occasionally these vessels would secure a freight of timber from the Baltic in summer or return with cargoes of scrap iron or saleable shingle to the North East coal ports. But the colliers, unlike the deep-water vessels, were far from being renowned for their quality. Two of them gained a well merited notoriety by being named in Samuel Plimsoll's 'Black List' of ships presented to the Board of Trade in 1873 in order to back his allegations of floating coffins and sea-going death-traps.

The *Friendship* was one of the big barges built at Sittingbourne in 1890 and loading 420 tons. She was eventually run down off Spurn Head in 1911 by a steam collier. *Amos*

The continuing rise in losses of men and ships and the conditions of payment and employment, nowhere more acute than in the coal fleet, had been a subject for concern and discussions for years before Samuel Plimsoll had been elected to Parliament in 1868. James Hall, a Tynemouth shipowner and insurer, with the problem of over-insured, under-maintained colliers at his door, had already in 1866 presented a memorial to the Board of Trade suggesting government inspectors of shipping at all the principal ports. All vessels, he proposed, should be subject to examination and if found wanting they should be refused permission to proceed to sea. The Board of Trade rejected the idea, but favoured a system which provided for the compulsory survey of any ship, but only in the case of the crew deeming it to be unseaworthy. Whether this was to be a majority decision or a unanimous one was not made clear. To any one aware of the realities of the problem, with owners sending rotten ships out fervently hoping they would not be seen again and of the over insurance of ancient and unseaworthy vessels, it would be difficult to imagine a more unsatisfactory solution.

When the Merchant Shipping Bill was introduced into the House in 1869 James Hall pointed out that it lacked the essence of all shipping reform, a means of determining a 'line of extreme loading'. With hindsight, as we read of the savage losses along the coast and the lonely tragedies fought out at sea, we might wonder why such a modest reform was not accepted. There was however, the problem of a Liberal administration in power seeking humanitarian changes, but rejecting government intervention. There were expert technical advisors who, as always, could be relied upon to supply mutually contradictory advice, while ship-owners argued that unless other countries did likewise our merchant vessels would be put at a crippling disadvantage; all these factors prevented progress towards legislation.

Gradually a combination of Hall's methodical marshalling of facts and support, culminating in the North Shields and Tynemouth Chamber of Commerce expressing support for a fixed load line, combined with Plimsoll's determined and flamboyant pressure, produced results. Finally, in 1875 the principle of a load line was adopted, but as a contemporary remarked the disc was painted on the hull by the owner at such a height "as to ensure its being well out of the water under any circumstance of loading." Even this concession was only obtained from the shipowners' interest by removing any reference in the Bill which might extend their liability to pay damages for death or injury of seamen. But Board of Trade courts of inquiry in to loss at sea was now mandatory and provided salutary reading and committees were formed to frame general rules relating to freeboard. At last, in 1890, the compulsory load-line was enacted; it might, in all justice, have become known as the Hall Mark of Safety rather than the Plimsoll line. But for many seamen the law came too late, by the nineties most of the under-manned, over-loaded colliers had passed away, the majority to a violent end.

Insurance, it is true, had acted as a self-regulating mechanism of good management in the coal-trade, ensuring that colliers were sea-worthy, their sails and spars well maintained and their masters competent. This had nevertheless been unable to stand up to the impact of sudden and unmatchable competition from steam, managed on a different scale. Many of the sailing colliers relied upon coverage from the Mutual Insurance Societies known as the 'Mutuals' or 'Clubs'. These flourished as well organised associations at Sunderland, Whitby, Whitstable and Shoreham, to mention only a few on the collier-coast.

They had their own carefully framed rules relating to the standard and quality of equipment, of sails and ground tackle. Trading areas were limited to certain times of the year and some notoriously dangerous harbours were actually barred as a result of hard-won experience and regulations relating to ballasting and free-board were laid down. Accidents that occurred to vessels trading outside the prescribed limits were uncovered. The 'club' had its own surveyor and much depended upon his professional skill and integrity. However, stories were told of how less than completely scrupulous owners would, under cover of darkness, bring warps, sails and cables, up to the required level before surveying was due by passing equipment between one vessel and another. One kedge and a spare fore-topsail, served, by this method, for two or three vessels. But in a small community such deception did not go undetected.

Masters, lacking official certificates, were examined by the associations and had to assure the unpaid 'Board of Directors' that they were competent to command a member's vessel. Membership to the 'Mutual' was obtained by paying an entrance fee of two or three per cent of the value of the proposed vessel into the group's communally held funds. The vessel was then entered into one of a scale of categories, dependent upon its age and tonnage. This contribution to the funds was an annual one and after it had reached above a certain level the surplus was returned. In the event of vessel's loss, or a claim entered for damage, known as a 'protest', a levy was made upon each member of the club, the contribution being proportional to the value of the ships he had covered by the organisation. In some 'Mutuals' the members proudly flew a flag when in harbour to indicate membership of the port's club; but when hoisted at sea it indicated that the vessel needed assistance and other members were expected to render help. In some clubs the residual value of a declared total loss became the property of the group, in others it was shared between the owner and the club. This led to unsung sagas of improvised salvage and ingenuity by expeditions led by the surveyor to refloat stranded craft. The 'Mutuals' roots in the local community meant that they could call upon reserves of good will to prevent pillage and to forestall the attention of wreckers should the disaster take place near the vessel's home-port. It was even known for expeditions to set out for foreign shores to undertake salvage.

More sophisticated forms of shipping insurance existed in the North East alongside the simply-run 'Mutuals' with their low over-head expenses. Collier-owners could, and did cover their property by becoming members of more than one 'club'. The low value of bulk cargoes such as coal, ranking with timber as the least valuable of all paying cargoes meant that colliery-owners did not have the same lever with which to force the exclusive use of properly insured and adequately surveyed ships to carry their cargoes. Many colliers made voyages for years without any insurance against loss, and these were by no means the least seaworthy.

It is just possible to understand the reluctance of shipowners to accept government insistence upon the institution of a statutory load line; that the rules relating to the use of navigation lights at sea took almost as much lobbying and discussion to formulate is unbelievable. The present pattern was not established internationally until 1889 and not universally adopted until nine years later. Yet by the sixties, during the transitory period between sail and steam in the coal trade, the situation had become critical. Steamers drove through fleets of colliers working to windward causing damage and confusion. As darkness fell the Thames estuary became a nightmare of criss-crossing vessels, fishing craft and screw-colliers pounding on their way. Until 1858 it was optional for sailing vessels to carry any lights at all and then, after that date, it was only a single lantern suspended at the bowsprit-end or lashed aloft at the mast-head.

A North Shields captain must be given the credit for the fundamental change to coloured red and green side-lights. Captain James Whitehead was master-owner of the brig *John and Mary* and had personal experience of the terror of hearing the thrash of an approaching screw-collier in the darkness and having no way of anticipating its course. Whitehead became determined to do all he could to bring official notice to his cause and gave up his command to further his fight for its acceptance. He addressed the Company of Ship Captains at Shields, pleaded his case before the brethren of Newcastle Trinity House. It was all to no avail. The captain boarded all incoming vessels in the Tyne to solicit support and signatures for his petition, but he met with little favourable reaction. Hardly surprisingly an appeal to the Admiralty met with even less response. But he had one sympathetic listener in W.S. Lindsey, M.P. for Tynemouth, who had some seafaring experience. He rapidly appreciated the advantages of the proposal and brought the matter before Parliament. Whitehead's perseverance was rewarded and in due course his suggestions were incorporated into the Rules for the Prevention of Collision at Sea.

An account book of an old coaster can tell a great deal of the last days of sail in the coal trade. Before me is the water-stained volume which relates the day-to-day expenses and records the modest profits of a sailing-collier of the eighteen sixties. In it the master of the *Jessie Annandale* entered in a fine flowing hand with, it must be admitted, some uncertainty of spelling, the evidence of his successful efforts to sail and manage his ship in the coal-trade for his Wivenhoe owners between 1866 and 1882.

The *Jessie Annandale* was launched in 1857 at Sunderland as a brigantine of 123 reg. tons, loading some 200 tons of coals. Tradition has it that she was bought to be owned on the East Coast from down Channel and when purchased had a reputation of being a leaky ship. This problem was rectified without much trouble or expense and she spent the rest of her working life owned on the Colne at Wivenhoe in Essex, managed by a series of conscientious captains and owned by a group of the local gentry and tradesmen. Her portrait shows her rigged with a Bentick boom on the square foresail, in true collier fashion, and rather unusually, a fore-topsail fitted with Cunningham's patent roller-reefing gear for by the late eighteen sixties, when the account book begins, double topsails were more usual. a new patent topsail is specifically mentioned at one point in the account, costing the owners £8.2.0d. It was claimed that it enabled a boy to reduce sail from the deck without going aloft, but there was a tendency for the arrangement to jamb, just when it was needed most, in a rising wind and night coming on. The usual remedy was for the mate to go aloft, cursing, and with the aid of hammer lubricate the machinery by percussion. The account book notes numerous visits aboard by a blacksmith and I wonder if his expert attentions were required to repair the topsail's iron roller-reefing gear when less skilled hands had proved ineffective.

The account-book opens in style with a heading "The First Voyage of the *Jessie Annandale,* commencing on the 16th May 1866" written with a fine flourish. This was to Sunderland in ballast, to load coals for the vessel's home port, but before leaving they completed their equipment by adding "shoe brush 2s.6d., 2 pudding Bags at 1s." and paid the same price for a bread sack. The first voyage earned only a modest profit of £2.8.2½d., the accounting is nothing if not detailed, but they were still covering the initial expenses of fitting out for the coal trade and had enrolled a carpenter in the crew, perhaps to complete work while the voyage was in progress. A second voyage with coals to Rochester, this time from Seaham, is more profitable and produced nearly nineteen pounds profit.

The wages of the crew, including the skipper, were paid by the voyage and are listed as, Master £6.5.0d., mate £3.15.0., man £3.5.0., lad £2.10.0., apprentice £1.15.0., and for assisting in working out the cargo the 'lad' received a well earned extra five shillings. The ship fed them all, and fed them well by the standards of the time. Ten years later the payment of the crew changes to a wage calculated on a monthly basis and they appear to have gained something by the change. For a seaman of the mid-nineteenth century clothing must have been a major item of expense. It took the effect of the First World War to make mass-produced, and therefore cheap, clothing available and the second to devise really effective inexpensive protective garments. An invoice, made out to the mate of a collier and dating from this time lists "1 Pilot Vest 5s.6d.", "1 Reefer £1.3s.6d.", "2 shirts 4s.8d.", "1 Handkerchief 2s.9d.", "2 Drawers 6s.2d.", the total representing two-thirds of his monthly earnings.

Within a year's trading the *Jessie Annandale* had carried twelve cargoes, showing a profit of between £2 and £20 on each; she had landed coals at Wivenhoe, Ipswich and Rochester and moreover, had the good fortune to avoid incurring any serious expenses through damage. Only a charge of £2 for repairs to the mainsail speaks of misadventure. The year included a voyage from Ipswich to Cork, for which they shipped an extra hand and returning to the East Coast, via Llanelly, they loaded Welsh coals for Faversham. At Queenstown they were reduced to buying coal for the galley; surely an expense that must have raised an eyebrow when the owner cast an eye over the accounts.

One of the last of the collier brigantines on her way down channel about 1910. *A. Pyner*

There are regular purchases nearly every voyage of a side of beef for which upwards of three pounds are paid, while fish, vegetables and on one occasion rhubarb, are regularly noted, and in fair quantities, together with fresh bread and flour. "Grocery etc." probably indicated the buying of tea, coffee and sugar. When the profits for two voyages are abnormally low it is noticeable that a "Bullock head" and "Bullock heart" appear under "Expenses" and replace "Beef". Other domestic items are "Pudding bag and table cloth," purchased for 3s.6d., a candlestick for 6d., while an entry in the balance sheet for a voyage to Rochester lists "Blacksmith and new tea pot at 2s.8d.".

The expense of three shillings noted as a charge for sending rope from Wivenhoe to Sunderland, where the master collected it for use on his ship, is interesting. Local chandlers who had shares in small vessels expected that they would be patronised when stores were bought — even if it meant sending them by rail to a distant port. One of the reasons for the survival of such small colliers as the *Jessie Annandale* was that share-holders looked for profit not only from their share in the earnings but from supplying the never-ending necessities of a sailing vessel from a new main-sheet to lamp-oil.

A freight fixed for London, whence they brought coals from Seaham could usually earn a profit while a voyage to Bruges with coal from Sunderland was only moderately successful. A carefully detailed page gives the cost of the extra expenses of a refit, undertaken at Wivenhoe during the month of January. This was a time when it was more profitable for the vessel to lie safely in the soft mud of the Colne than be abroad on the North Sea. Together with another refit later it brings home the expense of running a small sailing ship; including the skipper's pay for nine weeks, as he supervised and assisted in the work, with the crew paid off, the total expenses are £176. A great deal of coal had to be carried to balance that sum. Later on a new mainsail for £20 and a mainstay sail costing £6, indicate a further unavoidable charge. Only once is a new cable mentioned, some 30 fathoms, costing four pounds and as no anchor is mentioned we must suppose that it is the replacement for one found defective and not a loss through stress of weather.

The years bring voyages to Newcastle from London, where instead of ballast they loaded a chance cargo of iron, then to Lowestoft and Bridport, where beef is cheap, and to Fecamp and later Exmouth where the ship makes a record profit. For the half year of 1876 the ship clears a modest £62, which, one would guess represents for her owners a five per cent return on their investment. Occasionally paying cargoes for the return to the coal ports are secured instead of voyages in ballast earning nothing. There is a cargo of Cornish clay taken to "Dort" (Dortrecht) from Par and after landing Sunderland coals at Teignmouth they load iron at Newport for Caen. The book's last entries are of three consecutive voyages to

Exmouth from Sunderland and then three to her own port as the darker months begin.

These are the bare bones of a ship's work and her crew's life. There is no mention of plunging out of the Tyne behind a yawing tug and then the long days beating against a South Westerly that seems never-ending – until the voyage to the North begins. It does not tell of anchoring amongst a fleet in the Downs as an on-shore gale begins and of dragging vessels looming out of the darkness and driving rain. Left out are the days groping through fog, the master straining to catch the sound of distant surf, moisture dripping off the grey motionless canvas and steamers blaring somewhere ahead. There are no entries to explain how they cleared Orford Ness by a cable length, or were nearly cast on the Haile Sands after a foresail sheet parted in a smother of sleet. Behind such laboriously penned statements as "settled up this day and shared by owners £17.5.6." are all this and much more.

How the *Gallant* probably appeared when she was employed in the coal trade. Reconstructed from documentary and other contemporary sources. *Drawing by the Author*

APPENDIX I

INVENTORY OF THE GALLANT

To be sold by Auction

The good Brigantine called the Gallant, Built in 1788 remarkable strong and well Built, good Sailer Burthen 103 Tons by Register carries near 7 Keels of Coal at 9½ Feet Water and shifts without Ballast, is well found and fit for Sea immediately.

INVENTORY

No.	
1	Main Mast
1	Trysail Mast
1	Main Top Mast
1	Top Gallt Mast
1	Main Boom
1	Gaft and Yard
1	Main Yard
1	Topsail Yard
1	Fore Mast
1	Fore top Mast
1	Top Gallt Mast
1	Fore Yard
1	Foretop Sail Yard
1	Foretop Gallt Yard
1	Bowsprit
1	Spritsail Yard
1	Jib Boom
1	Spare lower Yard
1	Spare Top Mast
	All the Standing and Running Rigging
2	Spare Top Gallt Masts
5	Steering Sail Booms
3	Steering Sail Yards
3	Setting Booms
	Sails
1	Trysail
1	Square mainsail
2	Maintopsails
1	Maintopgallt Sail
2	Mainstay Sails
1	Middle Stay Sails
1	Foresail
1	Foretopsail
1	Foretop Gallt Sail
2	Fore Stay Sails
1	Jib
2	Topmast steering Sails
3	Boats Sails
2	Port Sails
	a parcel of old Canvas
	Anchors
2	Bowers
2	Kedges
1	Boats
	Cables Etc.
1	Best Bower 90ths
1	Small Bower 80ths
1	Towline 70
1	Small Towline 30
1	Chain Rope 10
2	Buoy Ropes
1	Nun Buoy
2	Pumps
2	pump Brakes
1	Pump Hook
1	Sounding Rod
5	Spare pump Boxes

1 Shot pump
4 Pump Bolts
7 Lower Boxes

In the Hold

3 Runner and 3 double Block, with the Geer for a heavy Purchase
1 Spanish Burton
1 pair of Bale Slings
1 pair of Canhooks
2 pair of Pipe Slings
Sundry pieces of old Rope
1 Spare Runner 6 ft. by 5½ in.
2 pair of Steer Sl Halliards
2 pair of Steer Tacks
3 Top Blocks
2 Spare Topgallt Caps
1 Bithead Stopper
1 Deck Stopper
2 Cathead Stopper
2 Shank Painters
Sundry Mats for the Yards, Rigging Etc. Sundry spare Blocks, Sheaves Etc.
1 Spunyarn Winch and Bolt
1 Pitch pot and Ladle
2 Tar Brushes
4 Iron bound water casks
3 Iron bound Beer casks
2 provision casks
3 Ballast) Shovels
2 Corn)
4 Ballast Baskets
2 Iron Crows
1 Boat
2 Boat Skids
2 Boats Masts
1 Boats Spreat
5 Boats Oars
2 Boats Hooks
1 Boats Tiller
2 Boats Chocks
1 pair of Gripes
1 Pair of Lashings
2 Boats Daivid's
1 Fire Hearth and Pan
1 Ladle and Tormentor
3 Bowls and Platters
1 Side Ladder
1 Hold Ladder
4 Hatch Tarpaulines
2 Deck) Buckets
1 Draw)
1 Cat Block
2 Fish Hooks and Runners
1 Spanish Windless
2 Cranks
1 Nut Screw
4 Hand-spikes
Bars and Locks for the Hatches
2 Deck Brushes, Mop, Grindstone and Trough

Cabbin

1 Brass stove Copper pan and Funnel
1 Logline and Reel
1 Copper Candlestick and save all for the Binnacle
1 Hand line
2 Hand leads
1 Deep Sea line
1 Deep Sea Lead
1 Speaking Trumpet
2 Brass Compass
2 Wood Compasses
1 Two hour Glass
2 Half Minute Glasses
1 Telescope
1 Chart of the Western Ocean
2 Horn Lanthorns
1 Glass Lanthorne
2 Saucepans
2 Baking-pans
1 Gridiron
1 Trivet

1	Frying Pan
1	Cheese Toaster
1	Candlestick
1	Tinder Box Etc.
2	Spits
1	pair of Steelyards and Weights
2	Side Skeed Bolts
3	Steering Sail Boom irons
5	Scrapers
4	Becket Irons
3	Bale Handhooks
1	Boat Chain
1	Pair of Ropes
1	Tunnel
1	Porpoise Grain
1	Beam Dog
	a parcel of pump leather Sundry Hooks and Thimbles Iron Bolts and Nails
1	Cabin Bell
2	Marline Spikes
2	Serving Mallets
2	Boards – 2 Cable Fids
1	Spare topmast Fid
3	Mouse Traps
	a parcel of Spunyarn, points and Gaskets
2	Ensigns, 1 jack and 1 Pendant
1	Coffee Mill
1	Tea-pot and Canister
3	Cups and saucers
1	Mustard Pot pepper Box and Salt Stand
6	Plates and 1 Dish
30	Glass Bottles
	about 6 lbs of Paint, Pots Etc.
3	Knives and Forks
1	Chair, 1 Table
1	Saw, 1 Axe, 1 Adze
1	Plane 1 Shaving Knife
3	Hammers 1 Auger
1	Pair of Pinchers
6	Gimblets
9	Chissels and Augers
1	Caulking Mallet
	and
4	Caulking Irons.

(The vessel, upon auction, attracted an opening bid of 650 guineas, and was finally sold to 'W. Billing of Lynn in Norfolk mariner' for the sum of 755 guineas).

APPENDIX II

Extract from Account Book of brigantine *Jessie Annandale,* built 1857, Sunderland. Registered at Salcombe, 1857, Colchester, 1866. Dimensions 87.5 x 21 x 11, registered owner Abraham Harvey, Wivenhoe, Essex. Reproduced by kind permission of Hervey Benham. Original spelling and computation have been preserved.

October 21, 1868

	£	s.	d.
Disbursements at Seaham			
Grocery Bill	2	1	8
Pototoes etc.		5	0
Ship Chandler	1	14	3½
Black Smith		8	6
Balast and duck dues	3	7	2
Triming cargo	1	12	0
Pilots	1	7	0
Lawrences		5	0
	11	0	7½

	£	s.	d.
Wages for the Voyage			
Master	6	0	0
Mate	3	15	0
Man	3	5	0
Man	3	5	0
Lad	1	1	0
Apprentice		6	0
	17	12	0

October 30

	£	s.	d.
Expenses in London			
Delivery etc.	8	12	0½
Waterman	1	0	0
Balast etc.	2	7	0
Factors Charges	2	9	10
Beef etc.	2	18	0
Grocerys		6	9½
	17	13	8

	£	s.	d.
Expenses in All at Seaham	11	0	7½
Expenses in London	17	13	8
Expenses in all	46	9	3½
Full amount of fright	68	18	3
Balance due to Ship	22	11	11½

November 25, 1868

	£	s.	d.
Expenses at Sunderland			
Balast Bill	2	14	2
Triming Ship	1	15	0
Blacksmith		2	0
River dues		8	10
Fitters Fee		5	0
Light Bill	1	4	1
Commission on fright etc.	3	19	6
Potatoes		6	2
Steam Boat in from Sea		18	0
Steam up the river		5	0
Ship Chandlers Bill	2	11	9
Grocery Bill	2	14	7
Water		1	6
Beefe Bill	2	2	6
Steam Boat to Sea		18	0
Allowances		5	0
	20	11	1

December 19, 1868

	£	s.	d.
Expences at Bruges			
In all at Bruges	25	10	2
Wages in all for voyage	19	3	2
Paid for insurance on the fright	2	6	0
Paid by owner at Sunderland	2	9	0
Expances in all for the voyage	69	19	5
Full amount of fright	89	15	0
	19	15	7
Omitted Pravious		16	2
Balance due to ship	18	19	5

GLOSSARY

Balinger	Small craft, common in late-medieval and Tudor times. Rigged with two masts, they carried a variety of sail-plans, probably employing a spritsail and lateen mizzen.
Beef Net	A bag of coarse netting into which the day's allocation of salt-beef was placed to be towed astern in order to wash out some of the preservative.
Billander	Originally, in the seventeenth century, a small inland or coastal vessel; by the eighteenth century it denoted a two-masted craft, square-rigged on the foremast and a main-mast fitted out with a substantial dipping lugsail. Obsolete by the mid-nineteenth century.
Breaker	Small keg containing fresh water and carried on deck.
Brig	Two-masted vessel, square-rigged on each mast, distinguished from the snow (q.v.) by carrying a larger fore and aft gaff mainsail extended by a boom.
Brigantine	A designation of a two-masted rig which changed during the period under review, but was usually associated with relatively small vessels. At the beginning of the eighteenth century it referred to a two-masted rig, both masts carrying square-sails and a large fore and aft sail on the after one, either lug or gaff. By the mid-nineteenth century it implied a square-rigged foremast and a fore and aft rigged main-mast with a gaff topsail, and staysails set between them.
Cannel	Bituminous coal.
Cant	To heel a vessel over moderately. Also used to describe the edge of an alongshire shelf of mud or sand.
Chaldron (or Chalder)	A measure of capacity, possibly derived from the Latin 'celdra' – a certain measure. The London coal chaldron between 1266–1497 was 17 cwt., 1498–1529 21 cwt., 1530–1635 26.5 cwt., 1635–1680 26.6 cwt., 1703–1713 28 cwt., and finally between 1818 and 1830 25.5 cwt. The Newcastle chaldron likewise began at 17 cwt., rose in 1498 to 21 cwt. until 1529 when it became 42 cwt. Between 1635 and 1713 it stood at 53 cwt. and finally 56 cwt. from 1818 until 1830.
Chine	The angle between the sides and bottom of a flat-floored vessel.
Coble	An open boat of peculiar construction and indigenous to the North East coast. Its main characteristics are a deep fore-foot, twin keels to assist in beaching and a wave-like sheer; between 20 and 30 feet in length.
Crayer	Derives from 16th century word for carrier; a Tudor merchant vessel of small tonnage, squaresail-rigged as opposed to its contemporary the cromster (q.v.).

Crocket (Coquet) An entry outwards, signifying that all customs requirements had been met. A corruption of the Latin term *quo quiettus est.*

Cromster A type of vessel that appears towards the end of the sixteenth century; from records they were probably akin to a hoy or large coaster and distinguished from crayers by carrying a large fore-and-aft rig.

Culm Anthracite.

Fitter Tyneside agent who employed the keelmen.

Hoveller Longshoreman who carries provisions and lends assistance to ships, an unofficial pilot.

Hoy Used to describe a small vessel regularly trading with general cargo between two ports. In the eighteenth century it implied a small single-masted vessel rigged with a fore and aft sail set from a gaff with appropriate head-sails.

Lodes Narrow, man-made waterways, connecting and draining into rivers in the Fenland and adjacent areas.

Maul Heavy mallet or hammer.

Pink A narrow-sterned vessel of the late seventeenth and eighteenth centuries with the upper planking of the hull drawn out over the rudder-head. Usually of moderate or light tonnage.

Port A harbour, space of water or coastal area in the United Kingdom appointed by Treasury as a Customs' port; it might vary considerably over a period, affecting statistical returns.

Pound Locks Perhaps the earliest use was on the Exeter Canal in 1564–6. Vertical-sided rectangular chambers of masonry, closed by swinging wooden gates fitted with sluices which enable the chambers to be filled. The dimensions of locks regulated the size of the vessels employed upon the navigation.

Quant A long pole shod with iron and used for propelling a vessel in waterways.

Registrary Port A port approved by the Commissioners of Customs and Excise for the registry of British ships under the Merchant Shipping Acts. Small ports might be the base of a considerable fleet but not be a Registrary Port.

Richmond Shilling This was a levy peculiar to Newcastle coal, originally imposed by Queen Elizabeth to compensate for unpaid dues. In 1677 it was granted to the Duke of Richmond, whose family held the rights until the government bought them for £400,000 in 1799. Abolished in 1831.

Score Score of Chaldrons, about 41 tons, hence scorage, an additional commission paid to the London buyer by the shipper, usually the master of the collier at the earlier period covered by this book.

Sheer The longitudinal curve of a ship's side or deck.

Sloop When used to refer to a small merchant vessel it implies a single-masted rig carrying a single gaff-sail and appropriate head-sails.

Snow A two-masted vessel which became distinguishable c. 1730; square-rigged on both and with a trysail mast (q.v.) fitted directly abaft the main. On this a fore and aft sail was carried, set on a gaff but without a boom.

Staithe Staging built out into a river to facilitate loading of vessels; also stores of coal on North-East rivers.

Swatchway A narrow and sometimes uncharted channel between sandbanks.

Taffrail The curved rail at the upper part of a ship's stern; frequently carved and ornamented.

Tens The Ten was originally 10 chaldrons or 1 keel of coal and was the basis for the shipment of coal by the Newcastle Hostmen in the first quarter of the seventeenth century.

Trow A flat-bottomed vessel, peculiar to the Severn and Bristol Channel, rigged in a variety of ways, ranging from a single squaresail to a full coasting ketch rig. Its essential characteristics were an open hold protected by side-cloths and a transom stern.

Trysail A small gaff-sail carried in place of, or in addition to, the lower plain sail on any mast. The name derives from 'try' meaning to 'lie head to wind'. Of particular use on the fore-mast of a brig or snow when it was set as the only alternative to bare poles in full gale conditions.

Wagon On the 'Newcastle Roads' (wooden wagonways) these commonly held 42 cwt. and had wooden bodies, sloping outwards from a wooden base. By the mid-eighteenth century they were constructed to contain 53 cwt. During the 1800s two planks were added and they then held 70–80 cwt. and were known as 'Black Wagons'. They were employed at Seaham Harbour as late as the nineteen-sixties.

Wales Thick planks going the whole length of the exterior of a ship's hull, usually at the waterline.

Warp Substantial ropes, usually of considerable elasticity, used to moor a vessel; or to manoeuvre by lines made fast to a quay or buoy and then hove in by capstan or winch.

Wey (of salt) 2 cwt.–3 cwt., varying in time and place.

Winchester Bushel A round bushel with a plain and even bottom, being 18½ inches wide throughout and 8 inches deep. One chaldron equalled 68 Winchester Bushels. One London Chaldron equalled 36 Winchester Bushels. Henry III levied a duty on coal of one farthing for two quarters of coal 'measured by the King's measure'; this was a quarter of 8 Winchester Bushels which, in 1266, were one cubic foot each.

BIBLIOGRAPHY AND SOURCES

The *Mariners' Mirror,* publication of the Society of Nautical Research, is abbreviated to M.M.

CHAPTERS 1, 3, and 4

Anderson, R. and R. C. — *The Sailing Ship,* 1947 edition.

Clowes, G. S. Laird. — *British Fishing and Coastal Craft,* Science Museum, 1932.

Crumlin Pedersen, O. — *The Viking ships of Roskilde. Aspects of the History of Wooden Shipbuilding,* National Maritime Museum, 1970.

Galloway, R. L. — *A History of Coal Mining in Great Britain,* 1882, reprint 1970.

Nef, J. U. — *The Rise of the British Coal Industry,* 1932.

Carson, E. — *The Ancient and Rightful Customs,* 1972.

Smith, R. — *Sea Coal for London,* 1961.

Salisbury, W. — *Early tonnage Measurement in England,* M.M. Vol. 54.

Smailes, A. E. — *North England,* 1960.

Willan, T. S. — *The English Coasting Trade 1600–1750,* 1938.

The Free Shipwrights of England, Tract published 1705, with map of Tyne. Bodleian Library.

CHAPTER 2

Atkinson, F. — *The Great Northern Coalfield 1700–1900,* 1966.

Fordyce, W. — *Chart of the River Tyne,* 1846.

McCord, N. — *The Impress Service in N.E. England during the Napoleonic War,* M.M. Vol. 54.

Viall, H. — *Tyne Keels,* M.M. Vol. 28.

CHAPTER 5

Hadfield, C. — *The Canals of South and South East England,* 1969.

Priestley, J. — *Historical Account of the Navigable Rivers, Canals, etc.,* 1831.

Rolt, L. T. C. — *The Inland Waterways of England,* 1950.

Salis, H. R. — *Chronology of Inland Navigation,* 1897.

Skempton, A. W. — *The Engineers of the English River Navigations 1620–1760,* Proceedings Newcomen Society, 1953.

Vine, P. A. — *London's Lost Route to the Sea,* 1965.

White. — *Directory of Suffolk,* 1844.

CHAPTERS 6 and 7

Ashton, T. S. and Sykes, J. — *The Coal Industry in the Eighteenth Century,* 1929.

Barton, P. — *The Alum Ships,* The Dalesman, March, 1969.

Barton, P. — *The Port of Stockton-on-Tees and its Creeks,* Maritime History No. 1, 1971.

Chapman, F. H. af. — *Architectura Navalis Mercatoria,* Stockholm, 1968. Magdeburg, 1958.

Defoe, D. — *A Tour through the Whole Island of Great Britain,* (Everyman Edition, 1927).

Hughes, E. — *North Country Life in the Eighteenth Century,* The North East, 1952.

Mantoux, P. — *The Industrial Revolution in the Eighteenth Century,* 1968.

Rocque, J. — *An exact Survey of the City of London . . . and the County near 10 miles round London,* 1746.

Steel — *Elements of Mastmaking, Sailmaking and Rigging,* 1794 edition.

Steers, J. A. — *The Sea Coast,* 1962.

Davis, R. — *The Rise of the English Shipping Industry,* 1961.

Groenewegan, G. — *Hollandsche Schepen,* 1789.

Humble, A. F. — *An Old Whitby Collier,* Annual Report Whitby Lit. and Phil. Soc., 1972.

Richardson, W. — *A Mariner of England,* first published 1908 reprint 1970.

CHAPTER 8

Chatterton, K. — *Ships and Ways of other Days,* 1913.

Geeson, N. T. and Sexton, R. T. — *H.M. Sloop Investigator,* M.M. Vol. 56.

Hakluyt Society. — *The Journals of Capt. James Cook,* Vol. 1 1955, Vol. 2 1961.

Preston, C. — *Captain James Cook, R.N., F.R.S. and Whitby,* 1965.

CHAPTER 9

Baldwin, C. E. — *The History and Development of the Port of Blyth,* 1929.

Bower, F. C. — *Ship Types of London,* 1938.

Casson, W. — *Plan of the Rivers Tyne and Wear . . . and Staiths Thereon* etc., 1804 Newcastle.

The Story of the Tyne Tugs, Smith's Dock Journal, 1933.

Invention of the Lifeboat, The Naval Chronicle Vol. VII, 1802.

Nautical Magazine, 1868.

Illustrated London News, 1861.

Price, J. — *To Merchants, Manufacturers, Ship Owners and Public in General,* Broad Sheet, Gateshead 1838.

Smith's Dock Journal, Vol. 11, 1929.

CHAPTER 10

Brand, J. — *History and Antiquities of Newcastle upon Tyne,* 1789.

Cooke, E. W. — *Shipping and Craft,* 1828. Reprint, 1969, with commentary by Roger Finch.

Cooper, E. R. — *East Coast Brigs,* M.M. Vol. 31, 1945.

Report of Select Committee on of the Port of London, 1796.

Hair, T. H. — *Sketches of the Coal Mines in Northumberland and Durham,* 1839.

Hartwell, R. M. — *The Industrial Revolution in England,* 1965.

Kipping, R. — *Rudimentary Treatise on Mastmaking and Rigging,* 1853.

Elementary Treatise on Sail and Sail Making, 1865.

Lever D'arcy — *The Young Sea Officers Sheet Anchor,* London, 1819.

Mayhew, H. — *London Labour and London Poor,* 1851.

Jones Rodwell, L. — *The Geography of the London River,* 1931.

CHAPTER 11

Dougan, D. — *The History of North East Shipbuilding,* 1968.

Anon. — *The First Iron Screw Collier,* Smith's Dock Journal, 1929.

Large, D. — *The Third Marquess of Londonderry and the End of the Regulation 1844–5,* Durham University Journal, December, 1958.

Smailes, A. E. — *North England,* 1960.

Taylor, A. J. — *The Third Marquess of Londonderry and the North East Coal Trade,* Durham University Journal, December, 1955.

A Map of the Railways of the Newcastle on Tyne Coalfield in 1812.

Martin, S. B. and McCord, N. — *The Steamship Bedlington 1841–54,* Maritime History Vol. 1 1971.

Illustrated London News, August, 1852.

1800 Report from the Parliamentary Committee on the State of the Coal Trade.

1829 Report from Committee of the Lords on the State of the Coal Trade.

CHAPTER 12

Bouquet, M. — *South Eastern Sail,* 1972.

Carr, F. C. C. — *Sailing Barges,* 2nd edition 1971.

Greenhill, B. — *The Merchant Schooners,* Vols. 1 and 2 first edition 1951, second edition (revised) 1969.

Illustrated London News, 1856.

Nautical Magazine, 1875 and 1883.

Runciman, W. — *Collier Brigs and their Sailors,* first published 1926, reprint 1971.

A gang of coal-whippers at work below London Bridge. *Author*

Index

Q

R

S

S

The topsail schooner *Commerce*, built at Stockton-on-Tees in 1846, lies on the beach at Cromer ready to discharge her cargo of coal. She was lost on the North Norfolk coast in 1876. (see Chapter 6)

R. Malster

Index of Vessels

(Sailing Colliers unless otherwise stated.)

Detail from a Prospect of Newcastle, one of a series of towns and ports drawn and engraved by S. and N.